Geopolymers and Composites

This book offers comprehensive insight into recent advances in geopolymer composites and emerging processing technologies such as 3D printing that offer promising application prospects in a wide range of industries.

- Covers novel applications of geopolymers and composites in industries such as fire retardation coatings, refractory materials, water treatment filters, and marine structures.
- Offers guidance on joint treatment of industrial waste acids and solid wastes based on geopolymer technologies.
- Describes energy consumption, carbon emissions, and costs for various compositions of geopolymers, which provide an effective basis for industrialization.
- Provides guidance for design and preparation of geopolymer products based on typical local wastes.

With topical coverage that will help readers make full use of local resources and promote the sustainable development of enterprises, this reference is aimed at those working with new materials for furnaces, construction and building, civil engineering, and water treatment, among others.

Huirong Le is currently Professor & Director of the Future Materials Research Center in the Future Laboratory of Tsinghua University. He is engaged in research of lightweight composites, biomimetic nanocomposites, multifunctional surface coatings, and digital manufacturing. Prof. Le graduated from the Department of Materials Science and Engineering at Tsinghua University in 1989 and received a PhD in March 1994. Prior to returning to Tsinghua University, he worked at the University of Surrey, then the University of Cambridge as a Postdoctoral Research Associate, then Senior Research Associate. He then held Lectureship at the University of Dundee, Associate Professorship at Plymouth University, and Professorship at the University of Derby. He has been a Fellow of the Institution of Materials Minerals and Mining (FIMMM) since 2018 and Fellow of the Royal Society of Arts, Manufacture and Commerce (FRSA) since 2022.

Kaibao Wang is Researcher in The Future Laboratory and affiliated with the Department of Mechanical Engineering at Tsinghua University. He is engaged in research of geopolymers and composite materials. Dr. Wang graduated from the University of Strathclyde with a PhD and was awarded the Outstanding Student of the Year and Doctoral Scholarship. He graduated from the University of Strathclyde with a bachelor's degree and received a master's from the University of Glasgow. His undergraduate project won "The 2012 Armourers & Brasier's Company Prize" in the UK.

Longyuan Li is Professor of Structural Engineering in the School of Engineering, Computing, and Mathematics at the University of Plymouth. His research interests cover the durability of reinforced concrete structures, geopolymer concrete, cold-formed steel structures, and fire safety of concrete materials. Prof. Li has published over 200 technical papers in SCI journals with Scopus h-index 45. He is Fellow of the Alexander von Humboldt Foundation (Germany), Fellow of the UK Higher Education Academy, and Fellow of the Institution of Structural Engineers (UK). Currently, Prof. Li is an editor of *Construction and Building Materials* journal and a member of the editorial boards of *Magazine of Concrete Research* and *Journal of Marine Engineering & Technology.*

Emerging Materials and Technologies

Series Editor: Boris I. Kharissov

The *Emerging Materials and Technologies* series is devoted to highlighting publications centered on emerging advanced materials and novel technologies. Attention is paid to those newly discovered or applied materials with potential to solve pressing societal problems and improve quality of life, corresponding to environmental protection, medicine, communications, energy, transportation, advanced manufacturing, and related areas.

The series takes into account that, under present strong demands for energy, material, and cost savings, as well as heavy contamination problems and worldwide pandemic conditions, the area of emerging materials and related scalable technologies is a highly interdisciplinary field, with the need for researchers, professionals, and academics across the spectrum of engineering and technological disciplines. The main objective of this book series is to attract more attention to these materials and technologies and invite conversation among the international R&D community.

Hydrogen Production, Storage, and Utilization
Technologies and Applications
Abbas Tcharkhtchi, Hamidreza Vanaei, Albert Lucas and Sedigheh Farzaneh

MXenes for Energy Storage Applications
Emerging Characteristics, Compositions, and Synthesis Methods
Muhammad Rafique, M. Bilal Tahir, and Saira Anwar

Multi-scale and Multifunctional Coatings and Interfaces for Tribological Contacts
Ajit Behera, Kuldeep K Saxena, Dipen Kumar Rajak and Shankar Sehgal

Nanotechnology in Green Energy Generation
Ahmed Thabet Mohamed

Thermo-Acoustics of Nanofluids and Transfer Processes
Shriram S. Sonawane and Manjakuppam Malika

Hybrid Nanofluids
Heat and Mass Transfer Processes
Shriram S. Sonawane and Manjakuppam Malika

Applications of Hybrid Nanofluids in Science and Engineering
Edited by A. K. Pandey, H. Upreti, O. D. Makinde, and A. J. Chamkha

Solar Cells Development and Fabrication
Edited by Shivani Dhall, Kapil Sood and Vinay Gupta

Geopolymers and Composites
Processing Technologies and Applications
Edited by Huirong Le, Kaibao Wang and Longyuan Li

For more information about this series, please visit: www.routledge.com/Emerging-Materials-and-Technologies/book-series/CRCEMT

Geopolymers and Composites

Processing Technologies and Applications

Edited by

Huirong Le, Kaibao Wang, and Longyuan Li

CRC Press
Taylor & Francis Group
Boca Raton London New York

CRC Press is an imprint of the
Taylor & Francis Group, an **informa** business

Designed cover image: Shutterstock

First edition published 2025
by CRC Press
2385 NW Executive Center Drive, Suite 320, Boca Raton FL 33431

and by CRC Press
4 Park Square, Milton Park, Abingdon, Oxon, OX14 4RN

CRC Press is an imprint of Taylor & Francis Group, LLC

ISBN: 978-1-032-87973-4 (hbk)
ISBN: 978-1-032-87977-2 (pbk)
ISBN: 978-1-003-53568-3 (ebk)

DOI: 10.1201/9781003535683

Typeset in Times
by KnowledgeWorks Global Ltd.

Contents

Preface

At present, some books on geopolymers have been published, which elaborate on the research progress of geopolymers curing mechanism, material composition, physicochemical analysis, applications, etc., mainly based on alkali excitation system, for the applications of building materials, heavy metal fixation, and other normal temperature conditions, while the research on acid excitation system is less, and the application of refractory materials and other high temperature conditions is basically a blank. Under the background of high-quality strategic development goals such as the low carbon manufacturing and environmental protection, the book focuses on the 3D printing technology of geopolymers, the development of geopolymer composites, the resource utilization of solid wastes, the treatment of industrial waste acids, and the application of geopolymers in the fields of refractory materials and sewage treatment, which is in line with the high-quality development direction of today's society. The topic of the book will help relevant enterprises to enhance their market competitiveness, make full use of local resources, promote sustainable development, and play a leading role in the transformation of local economic and social development.

The first chapter of the book describes the latest progress, advantages, and potential development directions of geopolymers. The second chapter elaborates on the preparation of geopolymer composites, including the effects of fibre reinforcement and additives on the properties of geopolymers. The third chapter describes in detail the effects of industrial waste material types, material ratio, and curing process on the mechanical properties. The fourth and fifth chapters describe the current research progress of 3D printing geopolymers, including the preparation of materials and the optimization of printing parameters. Chapters six to ten discuss about the applications and practical cases of geopolymers in building materials, refractory coatings, marine structures, waste acid treatment, water treatment filters, etc. and compare them with existing materials in terms of performance, economy, and environment.

List of Contributors

U. Johnson Alengaram
Universiti Malaya
Kuala Lumpur, Malaysia

Anton Frederik Becher
Cracow University of Technology
Cracow, Poland

Szymon Gądek
Cracow University of Technology
Cracow, Poland

Yanhong Jia
Tsinghua University
Beijing, China

Sirithan Jiemsirilers
Chulalongkorn University
Bangkok, Thailand

Kinga Korniejenko
Cracow University of Technology
Cracow, Poland

Barbara Kozub
Cracow University of Technology
Cracow, Poland

Huirong Le
Tsinghua University
Beijing, China

Longyuan Li
Plymouth University
Plymouth, United Kingdom

Janusz Mikuła
Cracow University of Technology
Cracow, Poland

Mohammad Rauf
Tsinghua University
Beijing, China

Kaibao Wang
Tsinghua University
Beijing, China

1 Recent Advances in Geopolymer Cement and Concrete

Longyuan Li

1.1 INTRODUCTION

The concrete industry faces significant challenges related to carbon dioxide (CO_2) emissions and climate change. The production of ordinary Portland cement (OPC), a key component of concrete, is responsible for around 8% of global CO_2 emissions due to the energy-intensive calcination process and the chemical decomposition of limestone, which releases CO_2. The high emissions contribute to global warming and climate change, making it imperative for the industry to seek sustainable alternatives. Innovations like geopolymer concrete (GPC), which uses industrial by-products and emits fewer greenhouse gases, present a promising solution to reduce the environmental footprint of concrete production. Transitioning to such sustainable practices is essential for mitigating climate change impacts and aligning with global environmental goals.

The development of geopolymers for producing construction and building materials has a rich and multifaceted history, spanning from ancient practices to cutting-edge scientific advancements. Early forms of geopolymer-like materials can be traced back to ancient civilisations such as the Egyptians and Mesopotamians, who utilised natural pozzolans (volcanic ash) and lime in their construction techniques [1]. These early innovations laid the groundwork for the eventual rediscovery and scientific exploration of geopolymer materials in the 20th century. The modern era of geopolymer research began in the 1970s when French materials scientist Joseph Davidovits coined the term 'geopolymer' [2]. He discovered that aluminosilicate materials could be activated with alkaline solutions to form binder systems with superior properties compared to OPC. This breakthrough led to the development of the first geopolymer binders, which were initially tested for various applications, including GPC and geopolymer mortar, specifically for fire-resistant materials and thermal insulation. In the 1990s and 2000s, researchers expanded their investigations to include industrial by-products such as fly ash (FA), ground granulated blast-furnace slag (GGBS), and metakaolin (MK) as source materials for geopolymer binders. The primary goal was to develop sustainable alternatives to OPC, thereby reducing carbon emissions and utilising industrial waste. Extensive laboratory research during this period demonstrated the superior mechanical and chemical

DOI: 10.1201/9781003535683-1

properties of GPC, such as high compressive strength, low shrinkage, and excellent resistance to acids and high temperatures. By the 2010s, the focus shifted towards scaling up production and commercialising GPC. Pilot projects and field applications were conducted to demonstrate the feasibility and advantages of GPC in real-world conditions. Notable projects included infrastructure repairs, precast concrete products, and eco-friendly construction initiatives. Today, ongoing research and development continue to optimise mix design, curing processes, and performance characteristics of GPC. Efforts are also being made to standardise testing methods and develop guidelines for large-scale adoption in the construction industry. Future research is focusing on enhancing the properties of GPC, exploring new source materials, and integrating advanced construction technologies such as 3D printing and smart materials.

To illustrate the research progress on GPC, Figures 1.1 and 1.2 present the number of publications by year and by country/territory, respectively, based on a Scopus search conducted on July 1, 2024, using the keywords 'geopolymer concrete' and 'alkali-activated concrete'. The data reveals a substantial increase in publications over time, particularly after 2012. India leads with nearly 2,000 documents, followed by China and Australia. The United States, Malaysia, and the United Kingdom also demonstrate significant research outputs, indicating a global distribution of research efforts in this field.

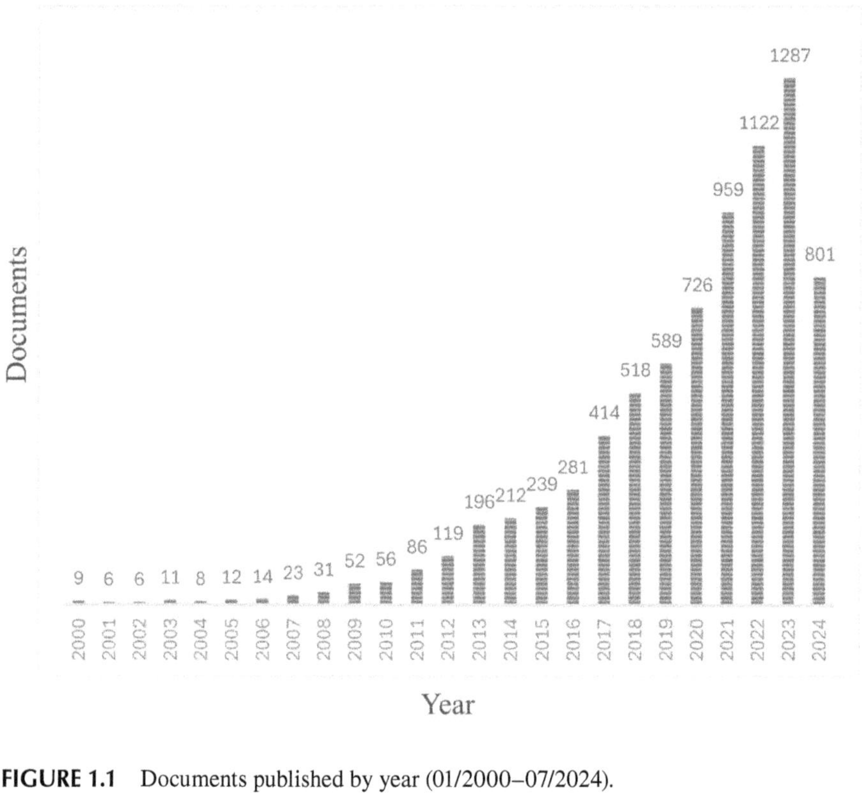

FIGURE 1.1 Documents published by year (01/2000–07/2024).

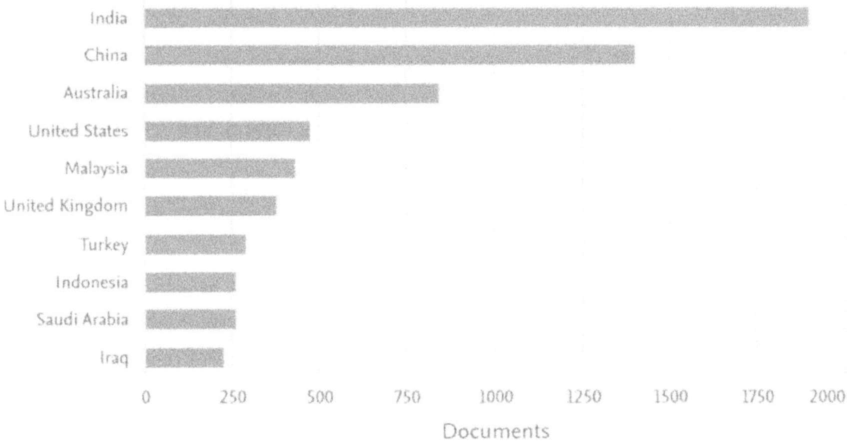

FIGURE 1.2 Documents published by country/territory (01/2000–07/2024).

1.2 GEOPOLYMER BINDERS

Geopolymer binders are created through the reaction of aluminosilicate precursors such as FA, GGBS, MK, and rice husk ash (RHA) with alkaline activators. The fundamental constituents of these materials include silicon (Si) and aluminium (Al), which undergo a series of chemical reactions during the geopolymerisation process. The primary steps involved in the synthesis of geopolymer binders are shown in Figure 1.3 and described as follows [3, 4]:

- Dissolution: The aluminosilicate particles break down in the presence of the alkaline solution, releasing Si and Al species into the solution.

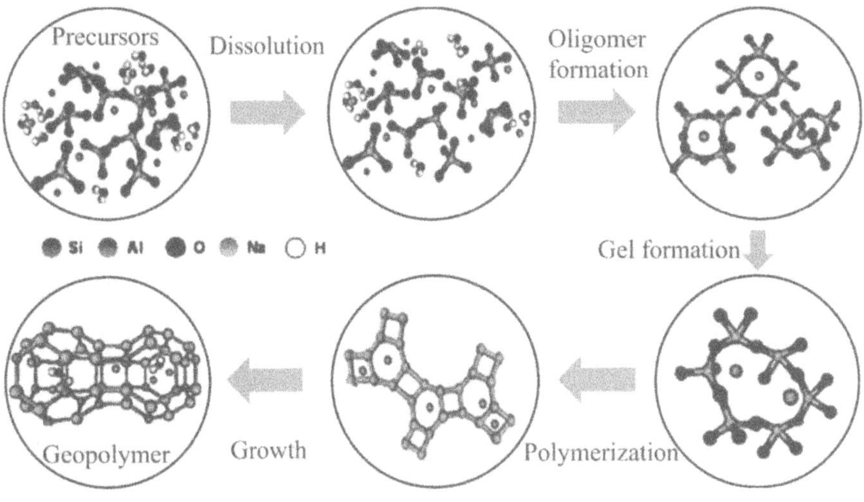

FIGURE 1.3 Dissolution and polymerisation processes of geopolymer precursors [4].

- Oligomer formation: The small molecules of Si and Al start linking together, forming small chains or clusters (oligomers).
- Gel formation: These oligomers grow larger and start forming a gel-like structure.
- Polymerisation: The gel structure becomes more rigid with the formation of a continuous 3D network.
- Hardened geopolymer: This is the final solid structure with an interconnected network of Si-O-Al bonds.

Geopolymer binders are typically derived from industrial by-products, such FA, GGBS, red mud (RM), and silica fume (SF), and/or from natural pozzolan materials, such as MK and RHA. Figure 1.4 shows the images of these materials [5]. The primary chemical constituents of these materials include Si and Al, which are crucial

FIGURE 1.4 Particle images of (a) FA, (b) GGBS, (c) MK, (d) RM, (e) SF, (f) RHA, (g) volcanic ash, (h) calcined clay, and (i) diatomaceous earth.

TABLE 1.1
Chemical Compounds Presented in Various Geopolymer Precursors (% by Weight)

Precursor	SiO_2	Al_2O_3	Fe_2O_3	CaO	MgO	Na_2O	K_2O	SO_3	TiO_2	P_2O_5	LOI*
Class-F FA	50–60	20–30	5–10	1–5	0–5	0.5–2.5	0.5–2.5	<1.5	<1	<1	0.5–5
Class-C FA	20–40	10–20	4–10	15–35	1–10	0.5–2.5	0.5–2.5	<1.5	<1	<1	0.5–5
GGBS	30–40	7–15	0.5–2.0	35–50	5–15	<1	<1	<1	<1	<1	<3
MK	50–55	40–45	1–2	<1	<1	<1	<1	<1	1–2	<1	<2
RHA	85–95	0.5–1.0	<1	<1	<1	<1	<1	<1	<1	1–5	<5
RM	5–20	10–30	30–60	2–8	<1	2–8	1–6	1–4	5–15	1–2	8–20
SF	85–98	0.5–1.5	<1	<1	<1	<1	<1	<1	<1	<1	<3
Natural pozzolans	50–60	15–25	5–15	1–10	1–5	0.5–2.5	0.5–2.5	<1.5	<2	<1	<5

* Note that loss on ignition (LOI) indicates the amount of weight loss during heating, often related to unburned carbon and other volatile components.

for the geopolymerisation process that forms the binder. Table 1.1 presents the primary chemical compounds typically found in various geopolymer precursors, along with their usual percentage ranges [5]. Note that the actual composition of these materials can vary depending on their source and processing methods. Although minor components like trace metals or other oxides are not listed in the table, they may still be present in small amounts.

FA is a fine, powdery material produced as a by-product during the combustion of pulverised coal in power plants. As coal burns, the non-combustible minerals within the coal (such as silica, alumina, and iron) melt and solidify into tiny, glassy particles that are carried out of the furnace with the flue gases. These particles are then collected using electrostatic precipitators or filter bags. FA is composed primarily of silicon dioxide (SiO_2), aluminium oxide (Al_2O_3), and iron oxide (Fe_2O_3), making it rich in the aluminosilicates necessary for the production of geopolymers. FA is typically classified into two main types, class F and class C, based on its chemical composition, which is influenced by the type of coal burned. Class F FA is produced from burning anthracite or bituminous coal. It contains low amounts of calcium oxide (CaO), making it rich in silica and alumina. Class F FA is pozzolanic, meaning it reacts with calcium hydroxide in the presence of water to form compounds with cementitious properties. Class C FA is derived from the combustion of lignite or sub-bituminous coal. It has a higher calcium content, often exceeding 20%. This higher CaO content gives it both pozzolanic and self-cementing properties, meaning it can set and harden on its own when mixed with water. FA-based geopolymers often exhibit superior compressive strength compared to traditional OPC, which makes them suitable for structural applications and excellent resistance to a wide range of chemicals, including acids, sulphates, and chlorides, which enhances their durability in harsh environments [6–8].

GGBS is a by-product obtained from the iron and steel-making industry. It is produced when molten iron slag, a non-metallic residue formed during the extraction of iron from its ore, is rapidly quenched with water or steam. This rapid cooling process forms a glazed granular material. The granules are then dried and ground into a fine powder known as GGBS. GGBS is primarily composed of CaO, SiO_2, Al_2O_3, and magnesium oxide (MgO), which gives it cementitious properties when mixed with water or alkaline activators. GGBS-based geopolymers develop high early compressive strength, making them ideal for applications requiring rapid strength gain. They also exhibit better workability and flow, low permeability, and enhanced resistance to harmful substances, extending the material's durability [9–12].

MK is a highly reactive pozzolanic material produced by the thermal treatment of kaolin, a natural clay mineral composed primarily of kaolinite. When kaolin is heated to temperatures between 650°C and 800°C, it undergoes a process known as calcination, which removes water from its crystal structure, resulting in the formation of MK. This calcined product is an amorphous, fine powder that is rich in silica (SiO_2) and alumina (Al_2O_3), making it an excellent precursor for the synthesis of geopolymers. MK-based geopolymers achieve high early strength, low shrinkage, and excellent resistance to chemicals, making them suitable for applications requiring rapid setting and durability. Their fine particle size and high reactivity improve workability and provide a smooth finish. Additionally, MK-based geopolymers offer outstanding thermal stability and low permeability, making them ideal for refractory and fire-resistant applications [13–15].

Rice husk is the outer protective covering of rice grains, which is removed during the milling process. It is a fibrous, tough, and abrasive material that constitutes about 20% of the weight of harvested rice. Rice husk is primarily composed of silica (SiO_2), lignin, cellulose, and hemicellulose. Due to its high silica content, rice husk is often used as a precursor for producing RHA through controlled burning. RHA is rich in amorphous silica, making it a valuable material for various industrial applications, including the production of geopolymer binders. RHA-based geopolymer binders are valued for their high reactivity and strength. These binders exhibit excellent durability, chemical resistance, low permeability, and thermal insulation, making them suitable for lightweight, energy-efficient construction [16, 17].

RM, also known as bauxite residue, is a by-product of the Bayer process, which is used to extract Al_2O_3 (alumina) from bauxite ore. The Bayer process involves treating bauxite with sodium hydroxide at high temperatures and pressures, which dissolves the alumina. The remaining residue, which consists primarily of iron oxides, alumina, silica, titanium dioxide, and various other oxides, is red in colour due to its high iron oxide content. RM is highly alkaline, with a pH ranging from 10 to 13, due to the residual sodium hydroxide from the Bayer process. RM-based GPC can achieve high compressive strength, comparable to or even exceeding that of conventional OPC concrete, especially when combined with other binders like FA and/or GGBS [18, 19].

SF, also known as microsilica, is a by-product of the production of Si metal or ferrosilicon alloys. It is produced in electric arc furnaces as SiO_2 vapour, which then condenses into extremely fine spherical particles. These particles are amorphous (non-crystalline) and have a very high content of SiO_2, typically around 85–98%. SF particles are extremely small, with an average diameter of about 0.1 μm, which is about 100 times smaller than the average cement particle. Due to its fine particle size, SF has a very high specific surface area, usually in the range of 15,000 to 30,000 m^2/kg. SF is a highly reactive pozzolan, meaning it reacts with calcium hydroxide to form additional calcium silicate hydrate (C-S-H), which contributes to the strength and durability of concrete. SF can be used as a source material of geopolymer binders, either on its own or in combination with other materials like FA, GGBS, or MK. Its fine particles and high reactivity enhance the properties of their produced GPC, making it a valuable additive in high-performance concrete applications [20, 21].

Natural pozzolans are naturally occurring materials that contain reactive silica and alumina, which, when finely ground and mixed with water, react with calcium hydroxide at ordinary temperatures to form compounds possessing cementitious properties. These materials have been used in construction for thousands of years, dating back to ancient civilisations like the Romans, who used volcanic ash in their concrete, contributing to the durability of structures like the Pantheon and aqueducts. The common types of natural pozzolans include volcanic ash, calcined clay, diatomaceous earth, and zeolites. Natural pozzolans can be used as a primary binder in GPC or as a supplementary material to enhance the properties of their produced GPC [22, 23].

It is important to note that the reactivity of the materials mentioned above as geopolymer precursors is influenced by factors such as particle size, crystallinity, specific surface area, and chemical composition. Figure 1.5 illustrates the particle sizes of various geopolymer precursors [24], which may directly impact the reactivity and effectiveness of the compounds listed in Table 1.1. Smaller particle sizes typically enhance the surface area available for chemical reactions, thereby increasing the reactivity of the compounds. Consequently, understanding the relationship between particle size and chemical composition is crucial for optimising the performance of geopolymer materials.

Note that many researchers use ternary diagrams to represent the components CaO, SiO_2, and Al_2O_3 in a material. In these diagrams, an equilateral triangle is depicted, with each corner representing 100% of one of the oxides (CaO, SiO_2, or Al_2O_3) [25]. Any point within the triangle corresponds to a specific combination of the three components, and its position indicates the relative proportions of CaO, SiO_2, and Al_2O_3 in the material. If the sum of CaO, SiO_2, and Al_2O_3 for a given

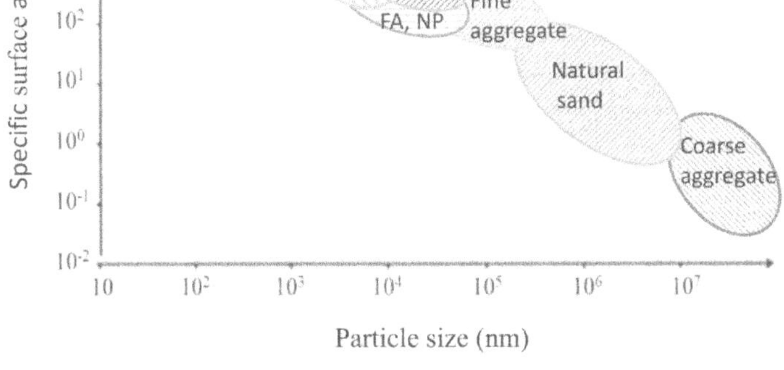

FIGURE 1.5 Particle size and specific surface area scale of FA, GGBS, MK, RHA, RM, SF, natural pozzolans (NP), OPC, sand, and fine and coarse aggregates [24].

material does not add up to 100%, the values must be normalised so that they in total give 100%. While normalisation allows for accurate plotting of individual materials on the ternary diagram, it can introduce inconsistencies when comparing different materials. This is because the normalisation process can obscure significant differences in the absolute content of the three oxides across materials, potentially leading to misleading interpretations if not carefully considered.

1.3 WORKABILITY AND MIX DESIGN OF GEOPOLYMER CONCRETE

1.3.1 Workability

GPC shares some similarities with traditional OPC concrete in its composition, consisting of solid components like the geopolymer precursor, sand, and coarse aggregate, and liquid components, primarily the alkaline activator. However, despite these similarities, the workability and overall behaviour of GPC can differ significantly due to the distinct chemical reactions involved in its formation.

Workability refers to the ease with which concrete can be mixed, placed, and compacted into its final form. In the context of GPC, workability is a crucial factor that directly impacts the efficiency of mixing, pouring, moulding, and compacting the concrete. Poor workability can lead to difficulties in placement, increased labour costs, and potential defects such as voids, honeycombing, or segregation. Achieving the right balance in workability is essential, as it needs to be aligned with other critical performance aspects, including strength, durability, and setting time. A mix that is too fluid may result in segregation and bleeding, while an overly stiff mix can be challenging to place and compact. Therefore, understanding and controlling the workability of GPC is vital for successful construction outcomes. Several factors influencing the workability of GPC can be listed as follows.

- Alkaline activators: The type and concentration of alkaline activators significantly affect workability. A higher concentration of activators can increase the viscosity of the mix, making it less workable. The choice between sodium hydroxide, sodium silicate, or a combination of both can alter the mix's fluidity.
- Source materials: The characteristics of aluminosilicate source materials play a crucial role. Fineness, particle shape, and chemical composition can affect the ease with which the concrete can be mixed and placed. Finer particles may improve workability by filling voids between aggregates, while coarser particles might reduce workability.
- Aggregates: The size, shape, and texture of aggregates influence the flowability of the mix. Rounded aggregates generally improve workability compared to angular aggregates. The moisture content of the aggregates also plays a role; wet aggregates can lead to unexpected variations in the liquid-to-binder (L/B) ratio, affecting the mix's consistency.
- Liquid-to-binder ratio: L/B ratio is a critical factor in determining workability. Higher L/B ratios typically increase workability by reducing viscosity, allowing the mix to flow more easily. However, excessive liquid can lead to segregation and bleeding.

- Aggregate-to-binder ratio (A/B): The proportion of aggregates to binder affects the density and viscosity of the mix. A higher aggregate content may reduce workability by increasing the overall stiffness of the concrete.
- Alkaline activator dosage: The dosage of alkaline activators, relative to the binder content, is crucial. Too much activator can lead to a rapid setting and a reduction in workability, while too little may not activate the geopolymerisation process effectively.
- Temperature: The ambient temperature during mixing and placing can significantly influence workability. Higher temperatures accelerate the geopolymerisation reaction, reducing the workability window. Conversely, lower temperatures may prolong setting time but improve workability.
- Curing conditions: The method and environment of curing (e.g., steam curing, ambient curing) affect the workability and final properties of GPC. Proper curing is essential to maintain the designed workability and prevent premature stiffening or cracking.

The workability of concrete is typically assessed using standard tests, each suited to different types of concrete and applications. For GPC, the most common tests include the slump test, flow table test, and Vebe test. Achieving the desired workability in GPC presents unique challenges due to the variability in source materials and the sensitivity of the geopolymerisation process to mix proportions and environmental conditions. GPC can set rapidly due to the high reactivity of alkaline activators, especially in warm climates. The use of retarders or adjustments in activator dosage can help control the setting time, improving workability. Additionally, admixtures like superplasticisers can significantly enhance the workability of GPC without compromising strength or durability. These admixtures reduce the viscosity of the mix, facilitating easier placement and compaction.

1.3.2 Mix Design

The mix design of GPC involves selecting and proportioning key materials such as geopolymer precursors, alkaline activators, and aggregates. The process focuses on balancing the binder, activator dosage, and aggregate ratios to achieve desired properties like workability, strength, and durability. Water content is minimised, with admixtures like superplasticisers added to enhance workability. Through trial batching and adjustments, the final mix is optimised to meet specific performance criteria, ensuring successful application in construction projects. Table 1.2 provides examples of mix design used for producing normal-strength, high-strength, self-compacting, low-heat, and rapid-setting GPC.

Once the mix design is established, the production of GPC involves a carefully controlled mixing process to ensure the optimal performance of the final product. The standard procedure for mixing GPC typically involves the following steps [26]:

- Preparation of materials: The aluminosilicate source materials (e.g., FA, MK) and the alkaline activator solution are prepared separately. The alkaline solution typically consists of a mixture of sodium or potassium hydroxide and silicate.

TABLE 1.2
Examples of Mix Design of GPC

Component	General-Purpose (30–35 MPa)	High Strength (50–60 MPa)	Self-Compacting	Low Heat	Rapid Setting
FA (kg/m³)	400	350	400	500	300
GGBS (kg/m³)	–	150	–	–	200
Sodium silicate solution (kg/m³)	100	120	120	80	130
Sodium hydroxide solution (kg/m³)	40 (12 M)	50 (14 M)	40 (10 M)	30 (8 M)	60 (16 M)
Sand (kg/m³)	650	600	850	700	650
Coarse aggregate (kg/m³)	1200 (10–20 mm)	1100 (10–20 mm)	900 (10 mm)	1100 (20 mm)	1150 (10–20 mm)
Water (kg/m³)	100	80	110	120	90
Superplasticiser (% of binder content)	1% (4 kg/m³)	1.5% (7.5 kg/m³)	2% (8 kg/m³)	1% (5 kg/m³)	1% (5 kg/m³)
Viscosity modifying agent (kg/m³)	–	–	0.2% (0.8 kg/m³)	–	–
Retarder (kg/m³)	–	–	–	–	0.2% (1 kg/m³)
Application	General structural components	High-performance elements	Complex forms, dense reinforcement	Mass concrete	Rapid repairs, emergency construction

- Mixing alkaline activator: The alkaline solution is prepared by dissolving solid hydroxide pellets in water and then adding the silicate solution. This mixture is allowed to cool to room temperature before use.
- Combining materials: The aluminosilicate source materials are dry-mixed with fine and coarse aggregates to ensure homogeneity. The prepared alkaline activator solution is then gradually added to the dry mix.
- Mixing process: The mixture is blended using a mechanical mixer to ensure a uniform distribution of the alkaline activator throughout the aluminosilicate material. This process typically takes several minutes, depending on the mixer type and batch size.
- Casting and compaction: The mixed GPC is poured into moulds and compacted to eliminate air voids. Compaction can be achieved through vibration or manual methods.
- Curing: GPC generally requires a curing period to achieve desired strength and durability. This can be done at ambient temperature or through heat curing, depending on the specific mix design and application requirements.

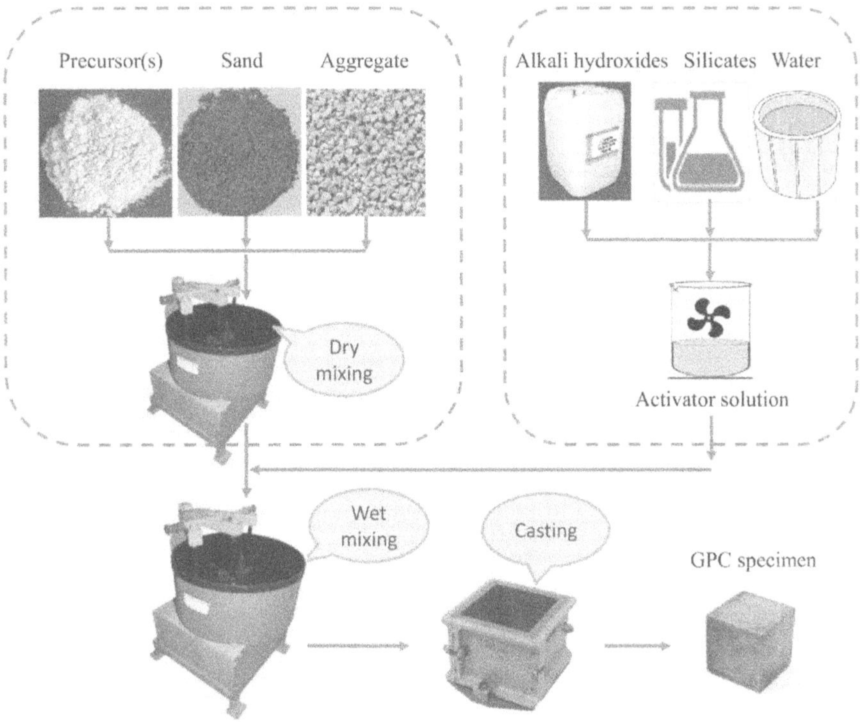

FIGURE 1.6 Standard procedure used for mixing GPC [26].

Note that the mixing process for GPC is more sensitive compared to traditional OPC concrete due to the critical nature of the alkaline solution's concentration. Adjusting workability in GPC involves managing the alkaline solution, a more complex task than adjusting water content and chemical admixtures in OPC concrete. Moreover, GPC often requires more stringent heat curing, whereas OPC concrete typically cures at ambient temperature with moisture retention methods. Safety considerations are also more pronounced in GPC due to the handling of corrosive alkaline activators, whereas OPC concrete, although requiring care, involves safer handling procedures. Figure 1.6 shows the standard procedure for mixing GPC.

1.3.3 Setting Time

The setting time of a binder is a crucial factor in its concrete application and performance, as it influences workability, curing requirements, and overall project timelines. The setting time dictates how long the material remains in a workable state, allowing it to be mixed, transported, placed, and finished.

Geopolymer binders and Portland cement exhibit notable differences in their setting times due to their distinct chemical compositions and hardening mechanisms. Portland cement sets and hardens primarily through hydration reactions, leading to the formation of C-S-H and calcium hydroxide. In contrast, geopolymer binders are set through geopolymerisation, involving the dissolution of aluminosilicate precursors in an alkaline solution followed by polycondensation reactions to form a hardened structure. The setting times for geopolymer binders can vary significantly, often exhibiting shorter initial and final setting times compared to Portland cement, which can be advantageous in certain construction scenarios but requires careful handling and precise control of the mixing and curing processes.

The following factors can influence the setting time of geopolymers.

- Alkaline activator concentration: The concentration and type of alkaline activator (e.g., sodium hydroxide, potassium hydroxide, or sodium silicate) play a pivotal role in controlling the setting time. Higher concentrations generally accelerate the dissolution of aluminosilicate precursors, leading to faster polycondensation reactions and shorter setting times. However, overly high concentrations can lead to rapid setting, reducing workability and potentially causing issues such as incomplete compaction or poor surface finish.
- Source material composition: The chemical and mineralogical composition of the aluminosilicate source materials, such as FA, MK, and BBGS, significantly affects the setting time. Materials with higher reactivity, such as MK, tend to react more rapidly with the alkaline activators, resulting in shorter setting times. Conversely, FA, which is less reactive, may prolong the setting time, allowing for longer workability periods.
- Curing temperature: Temperature is a critical factor in the setting and hardening process of geopolymer binders. Elevated curing temperatures can accelerate the polycondensation reactions, leading to faster setting and strength gain. This characteristic can be advantageous in situations where rapid strength development is desired. However, in ambient temperature curing, geopolymer binders might exhibit slower setting times, necessitating adjustments to the curing regime or the use of accelerators.
- Water-to-solid ratio: The water-to-solid ratio in the geopolymer mix influences its consistency and setting time. Lower water content generally leads to a faster setting time due to the higher concentration of reactive species in the solution, whereas higher water content can delay the setting time by diluting the reactants. Achieving the optimal water-to-solid ratio is essential for balancing workability and setting time.
- Additives and admixtures: The use of additives and admixtures, such as retarders, accelerators, or superplasticisers, can modify the setting time of geopolymer binders. Retarders can be used to prolong the setting time, providing more time for mixing and placing, especially in large-scale or complex pours. Accelerators, on the other hand, can reduce the setting time, which can be beneficial in cold weather conditions or when rapid turnover is required.
- Mix design and proportioning: The overall mix design, including the proportioning of source materials, activators, and additives, plays a critical role in determining the setting time. A well-balanced mix design that considers the specific requirements of the application, such as workability, setting time, and early strength, is essential for optimising the performance of geopolymer binders in concrete.

Similar to OPC, the setting time of geopolymer binders can be measured experimentally using a standardised Vicant test, in which a standard sample of the paste or mortar is prepared and placed in a cylindrical mould. A needle or plunger of a specified weight and size is then allowed to penetrate the sample at regular intervals. The initial setting time is recorded as the time elapsed from the start of mixing until the needle fails to penetrate the block to the 5.0 ± 0.5 mm from the bottom of the mould. The final setting time is the point at which the needle fails to make an impression on the surface of the test block.

Due to the variability in setting times, the use of geopolymer binders requires careful planning and control during the construction process. For instance, in precast concrete applications, where rapid setting and early strength development are often desired, geopolymer binders can offer significant advantages. However, in large pours or in hot weather conditions, managing the setting time is crucial to avoid premature stiffening of the mix, which can lead to challenges in placement and finishing. Moreover, the accelerated setting times associated with some geopolymer systems necessitate precise timing and coordination in the construction sequence. This might involve pre-planning the logistics of mixing, transportation, and placement to ensure that the material remains workable throughout the process. Additionally, the curing regime must be carefully controlled to prevent issues such as cracking or insufficient strength development, particularly in ambient temperature curing scenarios.

1.4 COMPRESSIVE STRENGTH AND STIFFNESS OF GEOPOLYMER CONCRETE

The compressive strength of GPC is a critical parameter that influences its structural performance and durability. The key factors that can affect the compressive strength of GPC include the chemical composition of precursors, L/B ratio, concentration of activator, A/B ratio, and curing conditions [5, 27–29].

1.4.1 Chemical Composition of Precursors

The chemical composition of the precursors used in GPC is fundamental to the geopolymerisation process and directly impacts the material's compressive strength. The precursors, typically sourced from industrial by-products like FA, GGBS, and MK, contain varying amounts of oxides such as silica (SiO_2), alumina (Al_2O_3), CaO, and Fe_2O_3. These oxides play distinct roles in the formation of the geopolymer matrix.

- Silica and alumina content: The ratio of silica to alumina is crucial in determining the strength and durability of the geopolymer matrix. A higher silica content typically enhances the formation of a strong aluminosilicate network, which increases compressive strength. An optimal silica-to-alumina ratio, often around 2:3, ensures that the matrix has adequate bonding and reduced porosity. A too-high silica content, however, may lead to an incomplete reaction or excessive silicate gel formation, which could trap unreacted particles and lower the overall strength.

- Calcium content: The incorporation of calcium-rich materials like GGBS into geopolymer mixtures introduces C-S-H gel formation alongside the primary aluminosilicate gel. The presence of C-S-H contributes to early strength gain and improves the overall compressive strength of the concrete. The synergistic effect of aluminosilicate and C-S-H gels creates a denser matrix with fewer microvoids. However, excessive calcium can lead to premature setting and increased brittleness, which may adversely affect the long-term strength and durability of the concrete.
- Iron oxide: Iron oxide acts as a flux in the geopolymerisation process, lowering the energy required for the reaction and enhancing the material's workability. The presence of iron oxide can improve the mechanical properties and density of the matrix, especially in the early stages. However, higher iron content might contribute to the formation of unwanted phases, such as ferrite, which can be less durable and reduce compressive strength over time.

1.4.2 Liquid-to-Binder Ratio

The L/B ratio is a critical parameter in GPC, which affects its workability, setting time, and compressive strength. This ratio determines the consistency of the mix and the degree of geopolymerisation.

- Higher L/B ratio: When the L/B ratio is high, the mixture tends to be more fluid, which can improve the workability and ease of casting. However, this increased fluidity often leads to higher porosity in the hardened concrete, as the excess liquid creates voids when it evaporates during curing. This porosity weakens the matrix, leading to a reduction in compressive strength. Additionally, too much activator relative to the binder can result in an incomplete reaction, leaving unreacted particles that further compromise strength.
- Lower L/B ratio: Decreasing the L/B ratio results in a thicker, less fluid mix, which reduces workability but leads to a denser and less porous matrix. This denser matrix is associated with higher compressive strength due to the reduced presence of voids and unreacted material. However, if the L/B ratio is too low, the mixture can become overly stiff, making it difficult to mix and place properly, potentially leading to poor compaction and weak spots within the concrete.

1.4.3 Concentration of Activator

The concentration of the activator solution, which typically consists of a combination of alkali hydroxides (e.g., sodium hydroxide or potassium hydroxide) and silicates (e.g., sodium silicate or potassium silicate), plays a pivotal role in the geopolymerisation process and the resulting compressive strength of the concrete.

- Alkali concentration: The concentration of alkali hydroxides is a major factor influencing the dissolution of silica and alumina from the precursor materials. A higher alkali concentration increases the pH of the solution, enhancing the dissolution rate and accelerating the geopolymerisation process. This generally results in higher compressive strength due to the more complete reaction and formation of a robust aluminosilicate network. However, very high alkali concentrations can cause rapid setting, leading to difficulties in handling and placing the concrete. Moreover, excessive alkali can lead to efflorescence and potential long-term durability issues, such as alkali-silica reaction (ASR).

- Silicate-to-hydroxide ratio: The ratio of silicate to hydroxide in the activator solution is another critical factor. A higher silicate content tends to promote the formation of a stronger, more coherent geopolymer matrix, thereby increasing compressive strength. The silicate acts as a binder that helps in the polymerisation of the dissolved alumina and silica. However, an overly high silicate content can increase the viscosity of the activator solution, making it difficult to mix uniformly with the binder. This could result in incomplete geopolymerisation and a reduction in the compressive strength of the hardened GPC.

1.4.4 Aggregate-to-Binder Ratio

The A/B ratio affects the density, homogeneity, and overall mechanical properties of GPC, including its compressive strength. The ratio determines the proportion of binder to aggregates, influencing the concrete's structural integrity.

- Higher A/B ratio: A higher aggregate content in relation to binder can lead to insufficient coating and bonding of the aggregates, resulting in a weaker concrete matrix. This scenario often leads to a higher porosity and a weaker interfacial transition zone (ITZ) between the binder and the aggregate, which is typically the weakest part of the concrete. The result is reduced compressive strength, as the poorly bonded aggregates are more likely to cause crack initiation and propagation under load.
- Lower A/B ratio: A lower aggregate content relative to binder generally results in a denser, more homogeneous concrete matrix, which enhances compressive strength. The decreased aggregate content improves the bonding between aggregates and reduces the presence of voids, leading to a more robust ITZ. However, an excessively high binder content can make the concrete more brittle and prone to cracking under load, which may limit its toughness and long-term durability. Additionally, a very high binder content can increase the cost and environmental footprint of the concrete.

1.4.5 Curing Conditions

Curing conditions, including temperature, duration, and humidity, are crucial in the development of the compressive strength of GPC. The geopolymerisation process is highly temperature-sensitive, and proper curing is essential to achieve optimal strength.

- Curing temperature: Elevated curing temperatures accelerate the geopolymerisation process by increasing the reaction rate of the precursors. Typically, temperatures between 60°C and 90°C are used to achieve high early compressive strength. This is particularly effective in promoting the formation of a dense, strong matrix in FA-based geopolymers, which benefit from higher activation energy. However, curing at excessively high temperatures can lead to rapid water loss and microcracking, especially if the curing is not properly controlled. This can result in a reduction of long-term compressive strength and durability.

- Curing duration: The duration of curing is directly related to the completeness of the geopolymerisation process. Longer curing durations allow for a more complete reaction, leading to a more stable and stronger matrix. In practice, a curing period of 24 to 48 hours at elevated temperatures is often sufficient to achieve the desired compressive strength. Beyond this period, the strength gain may plateau, especially if the reaction has consumed most of the available reactive components. However, insufficient curing time can leave unreacted particles in the matrix, resulting in lower strength and increased porosity.
- Humidity: Maintaining adequate humidity during the curing process is vital to prevent drying shrinkage, which can cause microcracks and reduce compressive strength. Humid curing environments help to retain moisture within the concrete, which is necessary for the continued geopolymerisation process. Inadequate humidity control, especially during the early stages of curing, can lead to surface cracking and a weaker overall matrix. In some cases, water curing or the use of curing compounds may be necessary to maintain the appropriate humidity levels.

For example, Figure 1.7 shows the variation of compressive strength of FA- and GGBS-based GPC with a L/B ratio [11, 30–32]. The results show a general decline in compressive strength as the L/B ratio increases. This trend suggests that higher L/B ratios, which introduce more liquid into the mixture, likely reduce the material's density and cohesion, leading to lower strength. However, it is important to note that other factors, such as the concentration of activators, sodium silicate-to-sodium hydroxide ratio, curing conditions, and the type of precursors used, can also influence the results, contributing to the variability observed in the data.

The relationship between compressive strength and the modulus of elasticity in GPC is similar to that of OPC concrete but with notable differences due to the distinct chemical composition of the two materials. In OPC concrete, the modulus of elasticity generally increases with compressive strength, following a well-established empirical relationship. However, in GPC, which is made from aluminosilicate materials activated by alkaline solutions, the modulus of elasticity tends to be lower for the same compressive strength compared to OPC concrete. This difference arises because GPC exhibits a more brittle behaviour and has a different internal microstructure, with less dense and more interconnected pore networks. As a result, while both materials show an increase in the modulus of elasticity with higher compressive strength, the rate of increase is typically lower in GPC, which can affect its overall stiffness and deformation characteristics. As an example, Figure 1.8 compares GPC and OPC concrete in terms of the relationship between compressive strength and the modulus of elasticity.

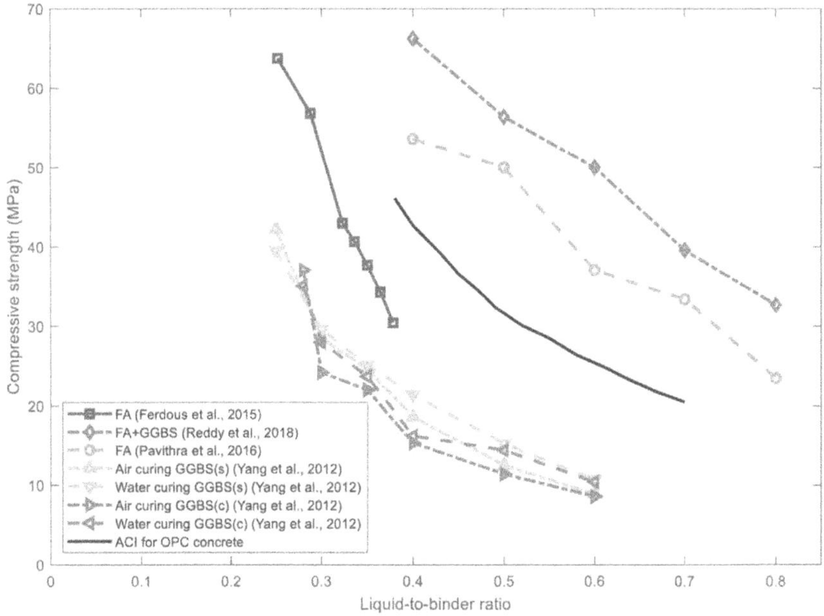

FIGURE 1.7 Effect of liquid-to-binder ratio on compressive strength of GPC. (Data taken from [11, 30–32].)

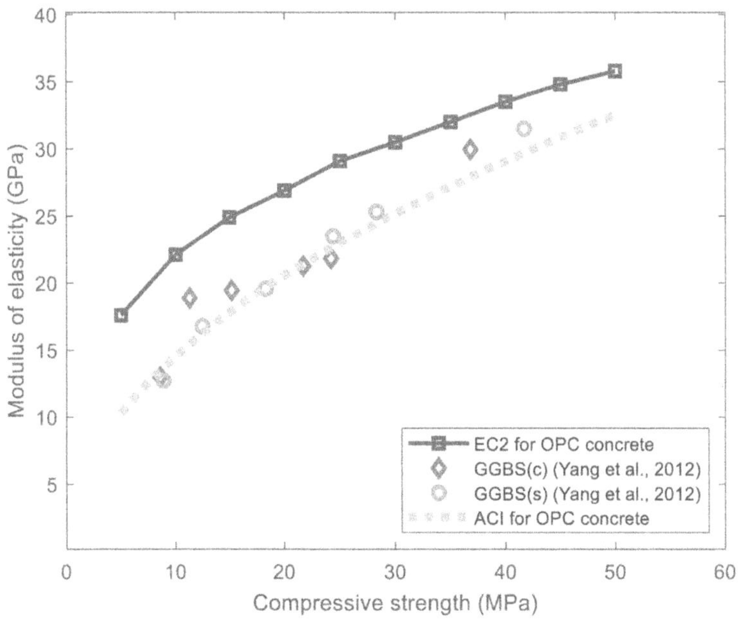

FIGURE 1.8 The relationship between compressive strength and the modulus of elasticity in GPC. (Data taken from [32].)

1.5 DURABILITY OF GEOPOLYMER CONCRETE

The durability of GPC is crucial because it directly impacts the longevity and sustainability of construction projects. Unlike traditional OPC concrete, GPC is highly resistant to chemical attacks, high temperatures, and environmental degradation, making it ideal for infrastructure in harsh conditions. By prioritising durability, we can reduce maintenance costs, extend the lifespan of structures, and contribute to more sustainable and resilient built environments. This section delves into the durability of GPC with a focus on its chemical resistance, freeze-thaw resistance, sulphate resistance, and abrasion resistance, providing a comprehensive understanding of its performance in challenging conditions.

1.5.1 Chemical Resistance

GPC is notably resistant to chemical attacks, a feature that stems from its unique chemical composition and microstructure. The primary binder in GPC is an aluminosilicate gel, which forms through the polycondensation of silicate and aluminate species in the presence of an alkaline activator, such as sodium hydroxide or potassium hydroxide. This process results in a dense, cross-linked network that is less reactive with aggressive chemicals compared to the C-S-H gel found in OPC concrete.

- Acid resistance: Traditional OPC concrete, with its high calcium content, is susceptible to acid attack. Acids react with calcium hydroxide and calcium silicate hydrate, leading to the dissolution of the binder and subsequent weakening of the concrete structure. In contrast, GPC's aluminosilicate gel is much less reactive to acids. The low calcium content in GPC reduces the availability of calcium hydroxide, which is prone to leaching in acidic environments. Moreover, the dense microstructure of GPC minimises the permeability of the concrete, limiting the ingress of acidic solutions. Studies have shown that GPC maintains its structural integrity and exhibits minimal mass loss even when exposed to highly acidic conditions (pH 3–4) over extended periods, making it ideal for use in environments such as industrial wastewater treatment plants and chemical processing facilities [33, 34].
- Chloride resistance: Chloride-induced corrosion of steel reinforcement is a significant issue in reinforced concrete structures, particularly in marine environments and regions where de-icing salts are used. Chloride ions penetrate the concrete, reach the steel reinforcement, and initiate corrosion, which can lead to cracking and spalling. The dense and low-permeability matrix of GPC significantly reduces the diffusion of chloride ions, thereby enhancing its resistance to chloride-induced corrosion. Research has shown that GPC has a chloride diffusion coefficient several times lower than that of OPC concrete, translating to a much longer service life in chloride-laden environments [35, 36].
- Alkali resistance: ASR is another durability issue in traditional concrete, where reactive silica in aggregates reacts with alkalis in the cement paste, leading to the formation of an expansive gel. This gel absorbs water, swells,

and causes cracking in the concrete. The aluminosilicate structure of GPC is less prone to ASR due to its different chemical composition and the absence of free alkalis in the hardened matrix [37, 38]. Furthermore, the lower permeability of GPC restricts the ingress of moisture, which is necessary for the expansion of the ASR gel, thereby enhancing its resistance to ASR.

1.5.2 Freeze-Thaw Resistance

Freeze-thaw cycles are a major durability concern for concrete structures in cold climates. When water within the pores of concrete freezes, it expands by approximately 9%, creating internal stresses that can lead to microcracking, scaling, and, eventually, significant structural damage. Traditional concrete mitigates this risk through the incorporation of air-entraining agents, which create small, evenly distributed air voids that provide space for the expanding ice.

- Low porosity and refined pore structure: GPC naturally exhibits a highly refined pore structure and significantly lower capillary porosity compared to OPC concrete. This is due to the geopolymerisation process, which results in a densely packed matrix with fewer and smaller capillary pores. The lower porosity of GPC reduces its water absorption capacity, minimising the amount of water available to freeze within the concrete. Consequently, the risk of freeze-thaw damage is significantly reduced.
- Absence of calcium hydroxide: In traditional concrete, the presence of calcium hydroxide can exacerbate freeze-thaw damage by facilitating the formation of ice lenses, the layers of ice that form within the pores of the concrete and cause additional internal stresses. GPC, however, contains little to no calcium hydroxide, which eliminates this risk. This absence of calcium hydroxide not only reduces the potential for ice lens formation but also contributes to the overall durability of GPC in freeze-thaw conditions.
- Experimental evidence: Laboratory studies and field tests have consistently shown that GPC outperforms OPC concrete in freeze-thaw durability. For example, GPC samples subjected to over 300 freeze-thaw cycles (in accordance with ASTM C666) have demonstrated negligible mass loss and minimal reduction in compressive strength [39]. This superior performance makes GPC an ideal material for use in cold climates, particularly for infrastructure that is frequently exposed to freeze-thaw cycles, such as bridges, pavements, and parking structures.

1.5.3 Sulphate Resistance

Sulphate attack is a common durability issue for concrete structures exposed to sulphate-rich soils, groundwater, or industrial effluents. In traditional OPC concrete, sulphate ions react with calcium hydroxide and the hydrated aluminate phases (such as monosulphate) in the cement paste to form expansive compounds like ettringite and gypsum. These compounds cause internal expansion, leading to cracking, spalling, and, ultimately, loss of structural integrity.

- Low calcium content: GPC is inherently resistant to sulphate attack due to its low calcium content. The absence of calcium hydroxide and calcium aluminate phases means that the chemical reactions that lead to the formation of expansive compounds do not occur in GPC. Instead, the aluminosilicate network that forms during the geopolymerisation process is highly resistant to sulphate ions, preventing the penetration and subsequent reactions that typically damage traditional concrete.
- Durability in aggressive environments: Field studies have shown that GPC performs exceptionally well in environments with high sulphate concentrations. For example, GPC specimens immersed in a 5% sodium sulphate solution for extended periods (over 12 months) exhibited minimal expansion and no significant loss of compressive strength [40]. This resistance to sulphate attack makes GPC an ideal choice for structures in sulphate-rich environments, such as sewage systems, underground foundations, and marine infrastructure.
- Resistance to combined chemical attacks: In many real-world scenarios, concrete structures are exposed to a combination of chemical attacks, such as sulphates, chlorides, and acids, all of which can simultaneously degrade the material. GPC's dense and chemically stable matrix provides a robust defence against these combined attacks, ensuring long-term durability in highly aggressive environments. This makes GPC particularly suitable for critical infrastructure in harsh chemical environments, where traditional concrete would rapidly deteriorate.

1.5.4 Abrasion Resistance

Abrasion resistance is a key durability factor for concrete surfaces that are subjected to mechanical wear and tear, such as pavements, industrial floors, hydraulic structures, and other high-traffic areas. Abrasion can lead to surface degradation, loss of material, and reduced functionality of the concrete structure.

- Dense microstructure: The abrasion resistance of GPC is largely attributed to its dense microstructure, which is a result of the geopolymerisation process. The aluminosilicate gel forms a tightly packed, three-dimensional network with minimal porosity and few microcracks. This dense microstructure provides a hard and durable surface that is resistant to mechanical wear. Additionally, the strong bonding within the aluminosilicate matrix ensures that the surface layer of GPC remains intact even under heavy abrasive forces.
- High compressive strength: GPC typically exhibits higher compressive strength than traditional OPC concrete, which contributes to its superior abrasion resistance. The higher the compressive strength of a concrete material, the more resistant it is to surface wear. GPC's high compressive strength is due to the strong chemical bonds formed during geopolymerisation, as well as the reduced presence of voids and microcracks. This results in a surface that can withstand heavy traffic, mechanical abrasion, and the impact of sharp or heavy objects without significant deterioration.

- Performance in abrasion testing: Standard abrasion tests, such as the ASTM C944 rotating cutter method and the Böhme grinding wheel test, have shown that GPC exhibits significantly lower material loss rates compared to OPC concrete under similar abrasive forces [41]. For example, GPC demonstrated an average wear depth of 2.8 and 4.4 mm after 12 and 24 hours of abrasion testing, respectively, whereas OPC concrete showed greater average wear depths of 4.5 and 7.2 mm at the same test intervals [42]. These results suggest that GPC surfaces are more durable and retain their integrity longer under abrasive conditions.
- Applications in high-wear environments: The superior abrasion resistance of GPC makes it particularly suitable for use in high-wear environments, such as industrial flooring in manufacturing plants, where heavy machinery, forklifts, and other equipment constantly exert mechanical stress on the concrete surface. It is also highly effective in transportation infrastructure, such as highways and airport runways, where the surface must withstand the constant wear and tear from vehicles and aircraft. In hydraulic structures, such as spillways and dams, GPC's abrasion resistance ensures long-term durability against the erosive forces of water and debris.

1.6 FIRE RESISTANCE OF GEOPOLYMER CONCRETE

The ability of construction materials to withstand high temperatures during fire events is a critical factor in ensuring the safety and longevity of structures. GPC, known for its superior durability and environmental benefits, has been shown to offer remarkable fire resistance compared to traditional OPC concrete.

1.6.1 Mechanical Properties of Geopolymer Concrete at Elevated Temperatures

Understanding how GPC behaves under fire conditions involves analysing its mechanical properties as it is exposed to elevated temperatures. These properties include compressive strength, tensile strength, modulus of elasticity, thermal conductivity, and thermal expansion, all of which are vital for maintaining structural integrity during a fire.

- Compressive strength: Compressive strength is a critical measure of concrete's ability to withstand loads without failing. At elevated temperatures, the performance of GPC in retaining its compressive strength is markedly superior to that of OPC concrete. This is largely due to the unique chemical structure of GPC, which is based on a network of aluminosilicate bonds rather than the C-S-H gel found in OPC concrete. When exposed to temperatures up to 200°C, GPC generally shows a slight increase in compressive strength. This phenomenon is attributed to the ongoing geopolymerisation and the densification of the microstructure as free water evaporates and additional bonds form within the matrix. Beyond this temperature, up to about 400°C, GPC begins to experience a gradual reduction in compressive strength. However, this reduction is far less severe than in OPC concrete, which can lose significant strength due to the dehydration and breakdown of C-S-H gel at similar temperatures. As temperatures rise to 600°C and above, GPC continues to outperform OPC concrete. At 600°C, GPC typically retains 70–80% of its original compressive strength, while OPC concrete may retain only 40–50%. This is due to the thermal stability of the aluminosilicate framework, which does not undergo the same level of chemical degradation as the C-S-H gel in OPC concrete. The absence of calcium hydroxide in GPC prevents the formation of lime, which decomposes at high temperatures in OPC concrete, leading to a loss of structural integrity. Figure 1.9 shows the typical variation of compressive strength with temperature obtained from experiments for the GPC mixed using FA, a combination of FA and GGBS, and GGBS. The data are taken from articles [43–47].

- Tensile strength: Tensile strength, the capacity of concrete to resist tension without cracking, is a critical factor in fire resistance, as it helps prevent the formation of thermal-induced cracks that can compromise the material's structural integrity. GPC shows a more gradual reduction in tensile strength as temperatures increase, compared to OPC concrete. At temperatures up to 200°C, GPC generally retains most of its tensile strength and may even see a minor improvement due to the same densification processes that enhance compressive strength. As temperatures approach 400°C, the tensile strength of GPC starts to decline, but this decline is less steep compared to OPC concrete. While OPC concrete can experience a sharp reduction in tensile strength due to the formation of microcracks from thermal expansion and the breakdown of the C-S-H gel, GPC's more stable matrix reduces the rate at which these microcracks form and propagate. At 600°C, GPC often retains around 60–70% of its tensile strength, compared to a significantly lower retention in OPC concrete. The aluminosilicate gel in GPC maintains its cohesive structure under thermal stress, which helps to resist crack formation and propagation, even at high temperatures.
- Modulus of elasticity: The modulus of elasticity, representing the material's stiffness, is another critical property that determines how much a material will deform under load. A high modulus of elasticity is desirable to ensure that the concrete can resist deformation during thermal expansion. GPC exhibits a slower reduction in modulus of elasticity compared to OPC concrete when exposed to elevated temperatures. At temperatures up to 400°C, GPC maintains a significant portion of its stiffness, whereas OPC concrete can experience a more rapid decline. The reason for this is twofold: firstly, the aluminosilicate matrix in GPC remains relatively intact, maintaining the bond strength between the particles; secondly, the low thermal conductivity and thermal expansion coefficient of GPC reduce the likelihood of internal stresses that cause microcracking. When exposed to temperatures as high as 600°C, GPC retains a higher percentage of its initial modulus of elasticity compared to OPC concrete. This means that GPC is less likely to deform excessively under thermal stress, maintaining its structural

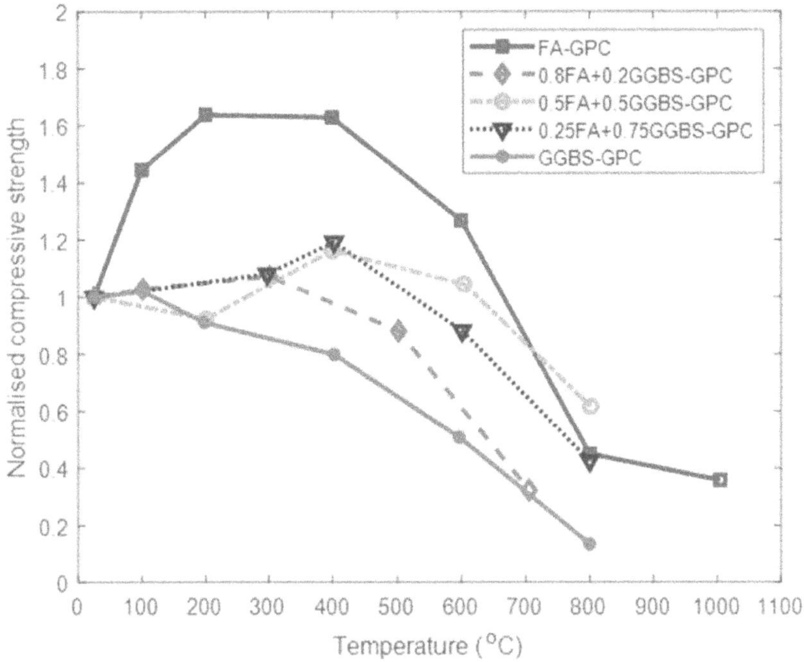

FIGURE 1.9 Normalised compressive strength of FA-, FA+GGBS-, and GGBS-GPC at elevated temperatures. (Data taken from [43–47].)

load-bearing capacity even in extreme conditions. This characteristic is particularly important in ensuring that structures remain stable during a fire, preventing excessive deflection or collapse. Figure 1.10 shows the typical stress-strain curves at different temperatures obtained from experiments for the GPC mixed using combined FA, GGBS, and SF. The data are taken from an article [48]. It can be seen from the figure that, with the increase in temperature, the compressive strength of GPC decreases while the strain at the peak stress point increases, indicating that GPC has much lower stiffness at higher temperatures.

- Thermal conductivity and thermal expansion: GPC's lower thermal conductivity compared to OPC concrete is a critical advantage in fire resistance. This lower conductivity means that heat penetrates the material more slowly, reducing the rate at which the temperature rises within the concrete's core. Consequently, GPC structures can withstand fire exposure for longer periods before reaching critical temperatures that could compromise their integrity. In addition, GPC has a lower coefficient of thermal expansion than OPC concrete. This means that GPC expands less when heated, which reduces the internal stresses caused by thermal gradients during a fire. These reduced thermal stresses minimise the risk of cracking and spalling, phenomena that are common in OPC concrete during rapid temperature increases.

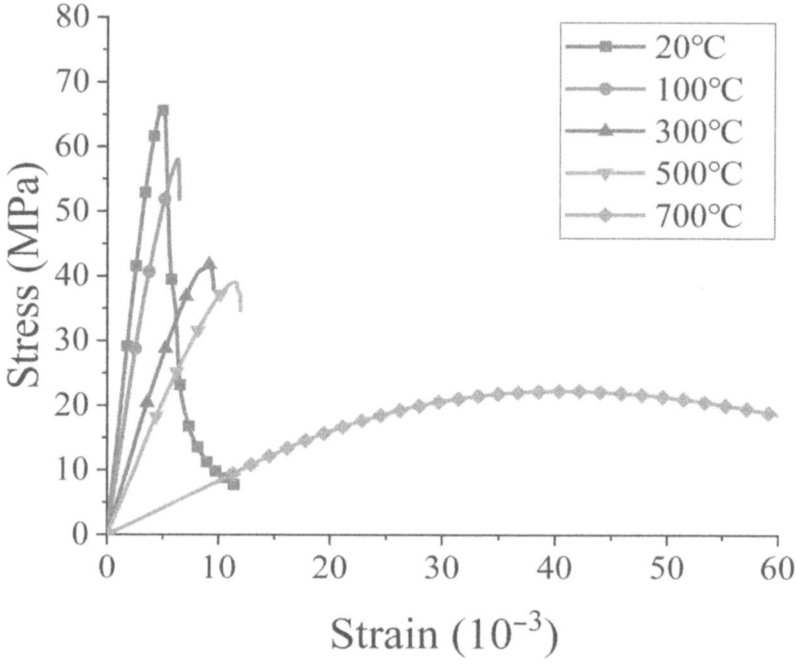

FIGURE 1.10 Experimental stress-strain curves of FA+GGBS+SF-GPC at elevated temperatures. (Data taken from [48].)

- Microstructural stability: The microstructural stability of GPC at elevated temperatures is one of its key advantages over OPC concrete. The aluminosilicate network in GPC remains largely intact even at temperatures up to 800°C, whereas the microstructure of OPC concrete can become severely compromised due to the dehydration of C-S-H gel and the breakdown of calcium hydroxide. The stability of the GPC matrix ensures that the material retains its integrity, with fewer microcracks and less porosity development compared to OPC concrete.

1.6.2 Residual Strength and Stiffness of Geopolymer Concrete after Exposure to Elevated Temperatures

The performance of GPC after it has been exposed to high temperatures is critical for assessing its viability for continued use or determining the extent of repairs needed. Residual properties, including compressive strength, tensile strength, and stiffness, are key indicators of the material's ability to function as a load-bearing element after a fire.

- Residual compressive strength: GPC exhibits superior residual compressive strength compared to OPC concrete after exposure to elevated temperatures. This is largely due to the thermal stability of the aluminosilicate

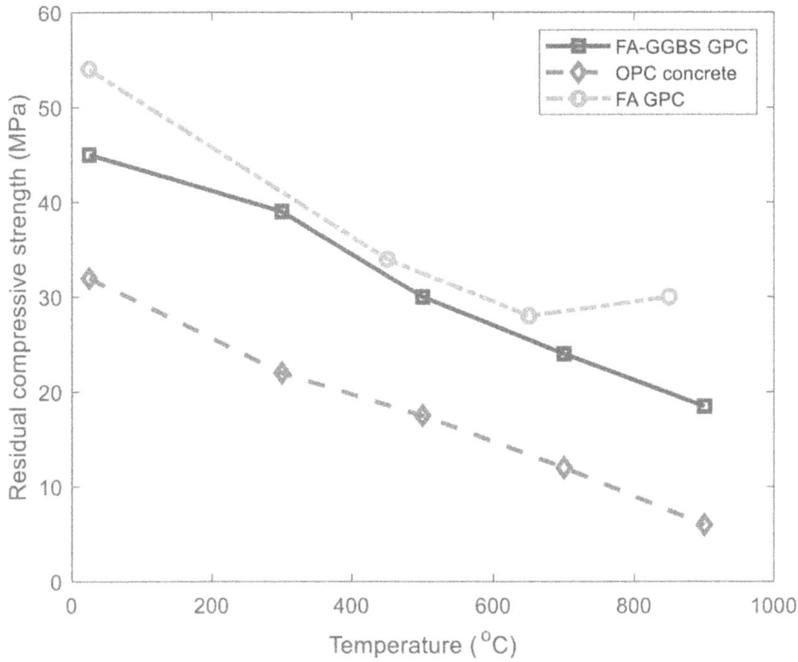

FIGURE 1.11 Residual compressive strength of OPC concrete and FA- and FA+GGBS-GPC after exposure to elevated temperatures. (Data taken from [51, 52].)

gel, which does not decompose or lose binding capacity as dramatically as C-S-H gel in OPC concrete. For instance, after being exposed to temperatures of 800°C and then cooled, GPC can retain 60–70% of its original compressive strength [49]. In contrast, OPC concrete typically retains only 30–40% of its strength after similar exposure [50]. This higher residual strength in GPC is critical for the continued safety of structures post-fire, as it means that the material can still bear significant loads without collapsing. The residual strength is also influenced by the minimal thermal cracking and spalling observed in GPC. Unlike OPC concrete, where fire exposure often leads to significant cracking due to thermal stress and the breakdown of hydration products, GPC's microstructure remains relatively intact, with fewer cracks and less damage overall. Figure 1.11 shows the variation of residual compressive strength with temperature obtained from experiments for the GPC mixed using FA and combined FA and GGBS. For the purpose of comparison, the results related to OPC concrete are also superimposed in the figure. The data are taken from articles [51, 52].

- Residual tensile strength: The residual tensile strength of GPC after exposure to high temperatures is similarly impressive. While all concrete experiences some loss of tensile strength after fire exposure, GPC typically retains a higher percentage of its original tensile strength than OPC concrete. This residual tensile strength is crucial in preventing further cracking and ensuring that the material can resist tensile forces in the structure. After exposure to temperatures around 600–800°C, GPC may retain up to 50–60% of its original tensile strength. The ability of GPC to maintain tensile capacity after cooling is largely due to the preservation of its microstructure and the minimal formation of thermal cracks, which would otherwise weaken the material's tensile performance.

- Residual modulus of elasticity: The residual modulus of elasticity, or stiffness, of GPC is another important property for post-fire structural assessment. GPC generally retains a significant portion of its stiffness after being exposed to elevated temperatures. This is crucial for maintaining the structural integrity of buildings, as a high residual modulus of elasticity ensures that the concrete can still resist deformation and maintain its shape under load. After exposure to temperatures up to 800°C, GPC often retains 50–60% of its original modulus of elasticity. In contrast, OPC concrete can suffer a much greater loss of stiffness due to the extensive microcracking and phase changes that occur within its matrix at high temperatures. The better retention of stiffness in GPC means that structures made from this material are less likely to experience excessive deflection or deformation after a fire, enhancing the safety and usability of the building. Figure 1.12 shows the typical residual stress-strain curves after exposed to different temperatures obtained from experiments for the GPC mixed using combined FA, GGBS, and SF. The data are taken from an article [53].
- Microstructural integrity and crack resistance: The microstructural analysis of GPC after exposure to elevated temperatures reveals that it suffers less microcracking and retains more of its original structure compared to OPC concrete. The aluminosilicate gel in GPC does not undergo significant chemical changes or decomposition at high temperatures unlike the C-S-H gel in OPC concrete, which can break down and lead to extensive cracking.

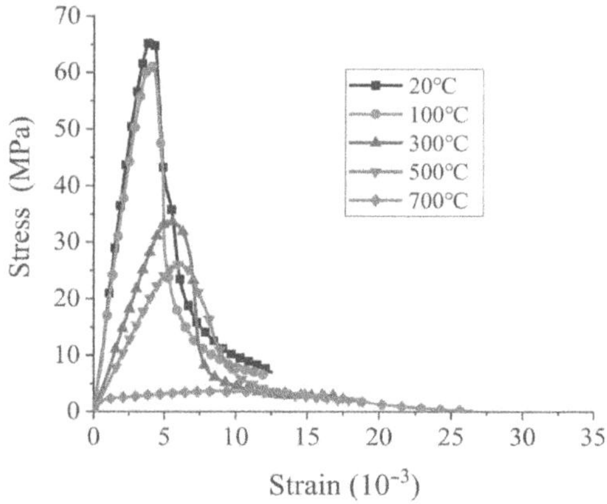

FIGURE 1.12 Experimental residual stress-strain curves of FA+GGBS+SF-GPC after exposure to elevated temperatures. (Data taken from [53].)

This microstructural integrity is crucial for the material's residual properties. Fewer microcracks mean that GPC retains more of its strength and stiffness, reducing the need for extensive repairs after a fire. The minimal cracking also limits the pathways for further damage from moisture ingress or chemical attack, which can occur in OPC concrete after fire exposure.

- Practical implications for structural safety: The superior residual properties of GPC after exposure to elevated temperatures have significant implications for the safety and resilience of structures. Buildings constructed with GPC are likely to exhibit fewer signs of structural damage after a fire, reducing the need for costly repairs or demolition. This resilience not only enhances the safety of the occupants but also contributes to the sustainability of the structure by extending its service life and reducing the need for material replacement.

In summary, GPC's excellent performance at elevated temperatures and its superior residual properties post-fire make it a highly attractive material for construction in fire-prone areas. Its ability to maintain strength, stiffness, and microstructural integrity under fire conditions underscores its potential as a robust and durable alternative to traditional OPC concrete.

1.7 CONCLUSIONS

GPC represents a sustainable and innovative alternative to traditional OPC concrete, addressing significant environmental concerns by reducing CO_2 emissions. GPC is produced through the activation of aluminosilicate materials using alkaline or phosphoric acid solutions. This process results in the formation of a robust binder that offers superior mechanical properties compared to OPC. The mix design of GPC is critical, requiring precise control over factors such as the L/B ratio, aggregate composition, and curing conditions to optimise performance. GPC demonstrates high compressive strength, comparable or superior to that of traditional OPC concrete. Its mechanical performance is influenced by the chemical composition of the precursors, the concentration of the alkaline activators, and curing conditions. Notably, GPC achieves high early strength, making it suitable for a wide range of structural applications. GPC exhibits excellent durability, particularly in harsh environments. It shows superior resistance to chemical attacks, including acid, sulphate, and chloride exposure, making it ideal for infrastructure in aggressive environments. Additionally, GPC has been found to possess significant freeze-thaw resistance and high abrasion resistance, further extending its service life in demanding conditions. One of the standout features of GPC is its fire resistance. Unlike OPC concrete, GPC retains a significant portion of its mechanical properties when exposed to elevated temperatures, making it a safer option for structures at risk of fire. This fire resistance, combined with its excellent residual strength after fire exposure, highlights GPC's potential as a reliable material in fire-prone areas.

Future research should focus on optimising the mix design and curing processes to further enhance the properties of GPC. Exploring new source materials, such as alternative industrial by-products and NP, could reduce costs and further minimise environmental impact. Additionally, integrating advanced technologies such as 3D printing and the development of smart GPC materials could open new avenues for innovative applications. Understanding the long-term performance of GPC in various environmental conditions and scaling up production for broader commercial use are also critical areas for future investigation. By advancing these research areas, GPC can be further developed as a sustainable, high-performance material for the construction industry, contributing to global efforts to reduce carbon emissions and improve the resilience of infrastructure.

REFERENCES

1. A. Lucas, Ancient Egyptian Materials & Industries (3rd edition), Edward Arnold (Publishers) LTD, London (1948, 1959).
2. J. Davidovits, Geopolymers, Journal of Thermal Analysis 37 (1991), 1633–1656.
3. K.Z. Farhan, M.A.M. Johari, R. Demirboğa, Assessment of important parameters involved in the synthesis of geopolymer composites: A review, Construction and Building Materials 264 (2020), 120276,
4. C.J. Shi, A.F. Jiménez, A. Palomo, New cements for the 21st century: The pursuit of an alternative to Portland cement, Cement and Concrete Research 41(7) (2011), 750–763.
5. P.L. Cong, Y.Q. Cheng, Advances in geopolymer materials: A comprehensive review, Journal of Traffic and Transportation Engineering (English Edition) 8(3) (2021), 283–314.
6. Q. Fu, W.R. Xu, X. Zhao, M.X. Bu, Q. Yuan, D.T. Niu, The microstructure and durability of fly ash-based geopolymer concrete: A review, Ceramics International 47(21) (2021), 29550–29566.
7. M. Amran, S. Debbarma, T. Ozbakkaloglu, Fly ash-based eco-friendly geopolymer concrete: A critical review of the long-term durability properties, Construction and Building Materials 270 (2021), 121857.
8. P. Zhang, Z. Gao, J. Wang, J.J. Guo, S.W. Hu, Y.F. Ling, Properties of fresh and hardened fly ash/slag based geopolymer concrete: A review, Journal of Cleaner Production 270 (2020), 122389.
9. J. Oluwafemi, O. Ofuyatan, A. Adedeji, D. Bankole, L. Justin, Reliability assessment of ground granulated blast furnace slag/cow bone ash- based geopolymer concrete, Journal of Building Engineering 64 (2023), 105620.
10. A. Singh, S.S. Bhadauria, M. Mudgal, S.S. Kushwah, Effect of alkali activator dosage on compressive and tensile strength of ground granulated blast furnace slag based geopolymer concrete, Canadian Journal of Civil Engineering 49(1) (2021), 73–82.
11. M.S. Reddy, P. Dinakar, B.H. Rao, Mix design development of fly ash and ground granulated blast furnace slag based geopolymer concrete, Journal of Building Engineering 20 (2018), 712–722.
12. A. Islam, U.J. Alengaram, M.Z. Jumaat, I.I. Bashar, S.M.A. Kabir, Engineering properties and carbon footprint of ground granulated blast-furnace slag-palm oil fuel ash-based structural geopolymer concrete, Construction and Building Materials 101(Part 1) (2015), 503–521.
13. A. Albidah, A.S. Alqarni, H. Abbas, T. Almusallam, Y. Al-Salloum, Behavior of metakaolin-based geopolymer concrete at ambient and elevated temperatures, Construction and Building Materials 317 (2022), 125910.
14. F.A. Shilar, S.V. Ganachari, V.B. Patil, K.S. Nisar, Evaluation of structural performances of metakaolin based geopolymer concrete, Journal of Materials Research and Technology 20 (2022), 3208–3228.

15. F.A. Shilar, S.V. Ganachari, V.B. Patil, I.N. Reddy, J. Shim, Preparation and validation of sustainable metakaolin based geopolymer concrete for structural application, Construction and Building Materials 371 (2023), 130688.
16. Saloni, Parveen, T.M. Pham, Y.Y. Lim, S.S. Pradhan, Jatin, J. Kumar, Performance of rice husk Ash-Based sustainable geopolymer concrete with Ultra-Fine slag and Corn cob ash, Construction and Building Materials 279 (2021), 122526.
17. Z.H. He, L.Y. Li, S.G. Du, Creep analysis of concrete containing rice husk ash, Cement and Concrete Composites 80 (2017), 190–199.
18. Q.W. Wang, S. Han, J.H. Yang, X.Y. Lin, M.Z. An, Preparation of geopolymer concrete with Bayer red mud and its reaction mechanism, Construction and Building Materials 409 (2023), 133730.
19. A. Kumar, T.J. Saravanan, K. Bisht, K.I.S.A. Kabeer, A review on the utilisation of red mud for the production of geopolymer and alkali activated concrete, Construction and Building Materials 302 (2021), 124170.
20. F.N. Okoye, J. Durgaprasad, N.B. Singh, Effect of silica fume on the mechanical properties of fly ash based-geopolymer concrete, Ceramics International 42(2, Part B) (2016), 3000–3006.
21. S.K. Das, S.M. Mustakim, A. Adesina, J. Mishra, T.S. Alomayri, H.S. Assaedi, C.R. Kaze, Fresh, strength and microstructure properties of geopolymer concrete incorporating lime and silica fume as replacement of fly ash, Journal of Building Engineering 32 (2020), 101780.
22. R.H. Haddad, O. Alshbuol, Production of geopolymer concrete using natural pozzolan: A parametric study, Construction and Building Materials 114 (2016), 699–707.
23. J.V.S. Metekong, C.R. Kaze, J.G. Deutou, P. Venyite, A. Nana, E. Kamseu, U.C. Melo, T.T. Tatietse, Evaluation of performances of volcanic-ash-laterite based blended geopolymer concretes: Mechanical properties and durability, Journal of Building Engineering 34 (2021), 101935.
24. K. Sobolev, M.F. Gutiérrez, How nanotechnology can change the concrete world, American Ceramic Society Bulletin 84(10) (2005), 14–18.
25. X. Yu, J. Shi, Z. He, C. Yalçınkaya, V. Revilla-Cuesta, O. Gencel, Review of the materials composition and performance evolution of green alkali-activated cementitious materials, Clean Technologies and Environmental Policy 25 (2023), 1439–1459.
26. A.M. Arif, M. Shariq, Use of geopolymer concrete for a cleaner and sustainable environment – A review of mechanical properties and microstructure, Journal of Cleaner Production 223 (2019), 704–728.
27. W.W. Huo, Z.D. Zhu, W. Chen, J. Zhang, Z.Z. Kang, S.Y. Pu, Y. Wan, Effect of synthesis parameters on the development of unconfined compressive strength of recycled waste concrete powder-based geopolymers, Construction and Building Materials 292 (2021), 123264.
28. C. Ng, U.J. Alengaram, L.S. Wong, K.H. Mo, M.Z. Jumaat, S. Ramesh, A review on microstructural study and compressive strength of geopolymer mortar, paste and concrete, Construction and Building Materials 186 (2018), 550–576.
29. A. Bouaissi, L.Y. Li, M.M.A.B. Abdullah, Q.B. Bui, Mechanical properties and microstructure analysis of FA-GGBS-HMNS based geopolymer concrete, Construction and Building Materials 210 (2019), 198–209.
30. W. Ferdous, A. Manalo, A. Khennane, O. Kayali, Geopolymer concrete-filled pultruded composite beams – Concrete mix design and application, Cement and Concrete Composites 58 (2015), 1–13.
31. P. Pavithra, M.S. Reddy, P. Dinakar, B.H. Rao, B.K. Satpathy, A.N. Mohanty, A mix design procedure for geopolymer concrete with fly ash, Journal of Cleaner Production 133 (2016), 117–125.
32. K.H. Yang, A.R. Cho, J.K. Song, Effect of water–binder ratio on the mechanical properties of calcium hydroxide-based alkali-activated slag concrete, Construction and Building Materials 29 (2012), 504–511.

33. P. Chindaprasirt, P. Jitsangiam, W. Chalee, U. Rattanasak, Case study of the application of pervious fly ash geopolymer concrete for neutralisation of acidic wastewater, Case Studies in Construction Materials 15 (2021), e00770.
34. P. Chindaprasirt, U. Rattanasak, Synthesis of porous alkali-activated materials for high-acidic wastewater treatment, Journal of Water Process Engineering 33 (2020), 101118.
35. T. Yang, X. Yao, Z.H. Zhang, Quantification of chloride diffusion in fly ash–slag-based geopolymers by X-ray fluorescence (XRF), Construction and Building Materials 69 (2014), 109–115.
36. N.S. Amorim Júnior, J.S.A. Neto, H.A. Santana, M.S. Cilla, D.V. Ribeiro, Durability and service life analysis of metakaolin-based geopolymer concretes with respect to chloride penetration using chloride migration test and corrosion potential, Construction and Building Materials 287 (2021), 122970.
37. T. Williamson, M.C.G. Juenger, The role of activating solution concentration on alkali–silica reaction in alkali-activated fly ash concrete, Cement and Concrete Research 83 (2016), 124–130.
38. H. Rashidian-Dezfouli, P.R. Rangaraju, Study on the effect of selected parameters on the alkali-silica reaction of aggregate in ground glass fiber and fly ash-based geopolymer mortars, Construction and Building Materials 271 (2021), 121549.
39. M. Liu, H. Liu, M. Hua, C. Chen, X. Wang, X. Guo, T. Ma, Potential for recycling metakaolin/slag-based geopolymer concrete of various strength levels in freeze-thaw conditions. Materials 2024, 17, 1944.
40. R. Manjunath, M.C. Narasimhan, K.M. Umesha, Studies on high performance alkali activated slag concrete mixes subjected to aggressive environments and sustained elevated temperatures, Construction and Building Materials 229 (2019), 116887.
41. S.M. Cheyad, A.N. Hilo, T.S. Al-Gasham, Comparing the abrasion resistance of conventional concrete and geopolymer samples, Materials Today: Proceedings 56 (Part 4) (2022), 1832–1839.
42. K. Ramujee, M. Potharaju, Abrasion resistance of geopolymer composites, Procedia Materials Science 6 (2014), 1961–1966.
43. W.L. Tu, M.Z. Zhang, Behaviour of alkali-activated concrete at elevated temperatures: A critical review, Cement and Concrete Composites 138 (2023), 104961.
44. H.G. Zhang, L. Li, C. Yuan, Q.Y. Wang, P.K. Sarker, X.S. Shi, Deterioration of ambient-cured and heat-cured fly ash geopolymer concrete by high temperature exposure and prediction of its residual compressive strength, Construction and Building Materials 262 (2020), 120924.
45. F.L. Qu, W.G. Li, Z. Tao, A. Castel, K.J. Wang, High temperature resistance of fly ash/GGBFS-based geopolymer mortar with load-induced damage, Materials and Structures 53 (2020), 1–21.
46. P. Behera, V. Baheti, J. Militky, S. Naeem, Microstructure and mechanical properties of carbon microfiber reinforced geopolymers at elevated temperatures, Construction and Building Materials 160 (2018), 733–743.
47. O.A. Abdulkareem, A.M.M. Al Bakri, H. Kamarudin, I.K. Nizar, A.A. Saif, Effects of elevated temperatures on the thermal behavior and mechanical performance of fly ash geopolymer paste, mortar and lightweight concrete, Construction and Building Materials 50 (2014), 377–387.
48. T. Wang, M. Yu, H.J. Lin, D.W. Li, L.Y. Li, Experimental investigation of mechanical properties of GGBS-FA-SF blended geopolymer concrete at elevated temperatures, Fire Safety Journal 146 (2024), 104156.
49. K.M. Klima, K. Schollbach, H.J.H. Brouwers, Q.L. Yu, Thermal and fire resistance of Class F fly ash based geopolymers – A review, Construction and Building Materials 323 (2022), 126529.

50. J.A. Purkiss, L.Y. Li, Fire Safety Engineering, Design of Structures (3rd edition), CRC Press, Oxford, 2013.
51. M. Wasim, T.D. Ngo, D. Law, A state-of-the-art review on the durability of geopolymer concrete for sustainable structures and infrastructure, Construction and Building Materials 291 (2021), 123381.
52. S. Luhar, S. Chaudhary, I. Luhar, Thermal resistance of fly ash based rubberised geopolymer concrete, Journal of Building Engineering 19 (2018), 420–428.
53. M. Yu, T. Wang, Y. Chi, D.W. Li, L.Y. Li, F.Y. Shi, Residual mechanical properties of GGBS-FA-SF blended geopolymer concrete after exposed to elevated temperatures, Construction and Building Materials 411 (2024), 134378.

2 Processing and Properties of Phosphoric Acid-Activated Metakaolin Geopolymer Composites

Huirong Le

2.1 INTRODUCTION

As reported in 1978 by French scientist Joseph Davidovits, naturally occurring aluminosilicates, such as kaolinite, could be transformed at ambient temperature, in a short time, into three-dimensional aluminosilicate mineral polymer [1, 2]. This has attracted a lot of attention since the discovery. Scientists have widely studied the reaction mechanisms and physical and chemical properties of the materials and have found more and more applications in building materials, refractories, and so on [3, 4]. The material is also used to consolidate industrial and nuclear wastes containing heavy metal or radiative elements, which makes it particularly useful in the field of environmental protection [5]. However, its insufficient mechanical properties and high-temperature stability have hindered its wider applications in industries [6].

Alkaline-activated geopolymers have been widely studied but suffer from insufficient thermal resistance due to the formation of glass phases containing alkaline elements. Compared with alkali-activated geopolymers, phosphoric acid-activated geopolymers have superior high-temperature properties without the interference of alkaline elements [7]. However, there are few studies on phosphoric acid-activated geopolymers so far. Unlike the rapid curing of alkali-activated geopolymers, the curing of phosphoric acid-activated geopolymers is relatively slow and usually requires to be carried out above room temperature, adding more cost to fabrication [8].

Potential applications of thermal-resistant geopolymer ceramics include refractory materials for metallurgy and ceramics industries, mould cores for the casting industry, etc. However, the existing geopolymer materials suffer from insufficient mechanical strength and thermal resistance. Fibre reinforcement is regarded as an effective method to improve both the mechanical and thermal resistance of ceramics. A considerable number of fibres are employed as reinforcing phases to improve the properties of geopolymers, including natural fibres like cotton fibre [9], wood fibre [10], hemp fibre [11], and so on [12, 13]. These fibres are only suitable for room temperature applications. There are also studies using ceramic fibres as reinforcing phases, such as silicon carbide, alumina, mullite, and basalt fibres [14, 15].

 DOI: 10.1201/9781003535683-2

Fang et al. [16] investigated the mechanical properties of geopolymer composites with BN-coated SiC whiskers at elevated temperatures. The composites exhibited improved high-temperature strength up to 900°C. Yan et al. [17] investigated the high-temperature mechanical properties of C_f-SiC_f reinforced geopolymer composites, which also showed improved high-temperature resistance. However, The addition of SiC whiskers or SiC fibres increases the cost of the materials significantly.

This chapter is focused on phosphoric acid-activated geopolymers because of their superior high-temperature performance [18, 19]. Previous studies indicated that the geopolymerization mechanism in the reaction system of phosphoric acid and metakaolinite is believed to be a "bonding reaction" among the P-O tetrahedral units of phosphoric acid solution, the Si-O tetrahedral units, and the Al-O layer of metakaolinite. So the polymeric structure of -P-O-Si-O-Al-O formed during geopolymerization constitutes the main building block of phosphoric acid-based metakaolin geopolymeric structure [20, 21]. Unlike alkaline-activated geopolymer based on the -Si-O-Al-O building block, the -P-O-Si-O-Al-O building block does not require alkaline elements to balance the charge and hence improves the thermal resistance.

Fibre reinforcement method can also be applied to overcome its higher probability of brittle failure due to insufficient mechanical strength and temperature-induced stresses under a large temperature gradient. Mullite short fibres are widely used as the reinforcing phase because of their good temperature resistance [22], low cost, and similar compositions to the geopolymer matrix material studied. It is expected that mullite fibres will have good compatibility with the geopolymer matrix. The processing and the effect of the fibres on the microstructure and thermal mechanical properties have been reported by Wei et al [23]. The interplay of composition and processing will be further elaborated. Finally, the scope for further research will be proposed.

2.2 MATERIAL AND METHODS

2.2.1 Raw Materials

Kaolin powder was purchased from Sigma-Aldrich Inc; the specific surface area is 10.2 m^2/g according to the supplier. The kaolin powder was calcined at 800–900°C for 1 h to convert into metakaolin. The chemical compositions of kaolin and metakaolin determined by XRF are shown in Table 2.1. Mullite fibres were short fibres sourced from Rayjohn Refractories Ltd (Deqing County, Zhejiang Province,

TABLE 2.1
Chemical Compositions of Kaolin and Metakaolin

	Chemical Composition (wt%)								
Raw Materials	SiO_2	Al_2O_3	K_2O	Fe_2O_3	TiO_2	PbO	P_2O_5	BaO	MgO
Kaolin	54.37	39.78	2.11	0.916	0.662	0.602	0.445	0.200	0.197
Metakaolin	52.97	41.09	2.08	0.897	0.646	0.591	0.504	0.265	0.218

China). The average mullite fibre diameter is 3–5 μm, the average length is 2 mm, the alumina content is about 75%, and the melting point is about 1850°C. Tap water was used with phosphoric acid (85%) with a view for industrial applications in the experiments.

2.2.2 Sample Preparation

2.2.2.1 Fibre Dispersion

It is a widely adopted method to use short fibres as a reinforcing phase for a wide range of materials [24], but fibre dispersion remains a common problem to be solved. If the fibres are agglomerated and knotted, it is easy to generate defects in the composite and reduce the mechanical properties. The as-received mullite fibres are in the form of clusters. The initial idea to mix the fibres directly with kaolin did not break the agglomerates of the fibres. After many explorations, we developed an effective process to disperse the fibres.

The first step of fibre dispersion was to mix a certain amount of fibre with water and stir it quickly in the same direction. In this study, typically, the fibre weight was 5–20 g in 300 ml of water. A magnetic stirrer was used in this step in the laboratory. After about 30 min, a uniformly dispersed fibre suspension was obtained, and then a certain amount of kaolin powder was gradually added to the fibre suspension under continuous stirring. Finally, a uniform slurry of the fibre, kaolin, and water was obtained. The slurry was filtered by vacuum-assisted filtering to avoid agglomeration of fibres during the process, and then the obtained cake was dried in the oven and broken by dry ball milling. The milling time shall not be too long; typically, ball milling is at 150 rpm for 10 min and then at 300 rpm for 10 min to ensure the mullite fibres are not broken. Finally, the powder was calcined at 800–900°C [25, 26] for 1 h to obtain a mixture containing metakaolin and mullite fibres.

It was reported previously that when the molar ratio of Al_2O_3:H_3PO_4 = 1:1.3, phosphate-based geopolymer yields the best strength [18]. The recommended weight ratio of metakaolin to liquid phase is about 1:1 for ease of processing. So, phosphoric acid was diluted to a specific concentration and added to the powder as the liquid phase to ensure the desirable Al_2O_3:H_3PO_4 ratio. Since fibre is generally considered to be the part that does not participate in the reaction, phosphoric acid is added in proportion to the amount of metakaolin only. The compositions of the composites with various fibre contents are summarized in Table 2.2.

TABLE 2.2
Proportion of Raw Materials Used in Sample Preparation

Group	Metakaolinite (g)	Phosphoric Acid (85%) (g)	Water (g)	Mullite Fibre (g)
1	100	54	46	0
2	100	54	46	5
3	100	54	46	10
4	100	54	46	20

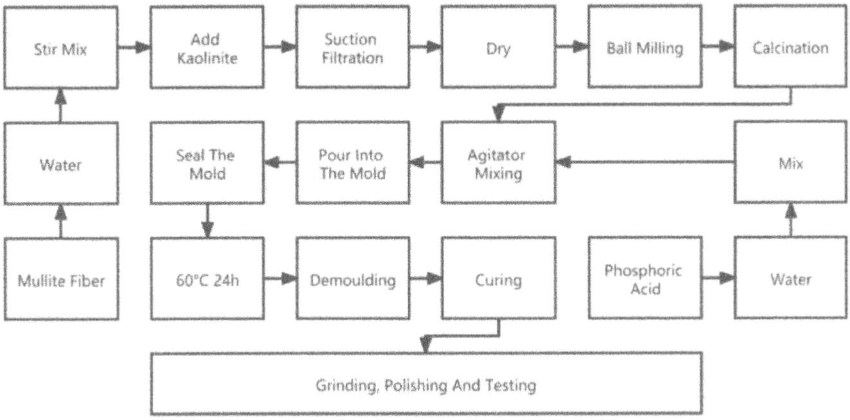

FIGURE 2.1 Preparation process of mullite fibre geopolymer [23].

2.2.2.2 Geopolymer Composite Preparation

The obtained solid and liquid phases are mixed in a mixer (MSK-SFM-7; Kejing, China) for 30 min and then poured into a plastic mould made by 3D printing, sealed, and placed on the shaking table at 120 rpm for 1 h to remove bubbles. The materials were cured for 24 h in an autoclave at 60°C and then removed from the mould and kept in humidity above 80% at room temperature for 7/28 days. Two types of samples were made for three-point bending and compression tests to obtain the flexural strength and compressive strength, respectively. The process flow chart is shown in Figure 2.1.

2.2.3 Microstructural Analysis and Mechanical Testing

Three-point bending and compression test samples are shown in Figure 2.2. Cylindrical specimens were used for the compressive strength test, while rectangular specimens were used for the three-point bending test, with a crosshead speed

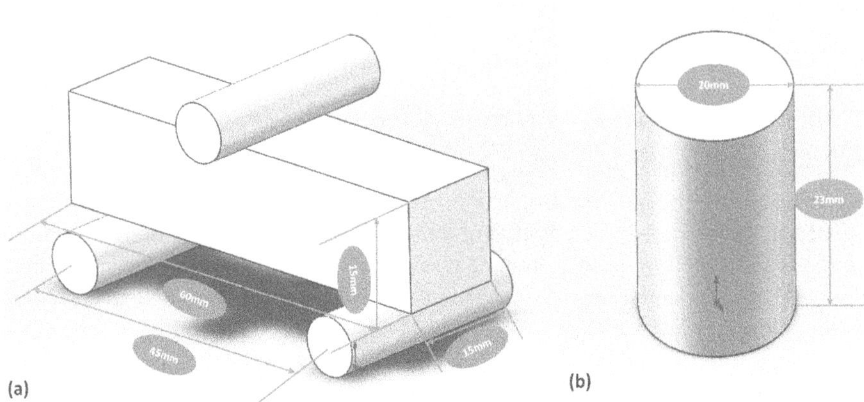

FIGURE 2.2 Specimen size for mechanical property test: (a) configuration of three-point bending, and (b) compressive strength test sample dimensions [23].

of 1 mm/min using a mechanical testing machine (50ST; Tinius Olsen, USA). Each group included three samples. As shown in Figure 2.2, the size of the rectangular sample for three-point bending was 60 mm × 15 mm × 15 mm, and the sample for the compression test was a cylinder with a diameter of 20 mm and a height of 23 mm. The peak load of six parallel test pieces was used to calculate the flexural strength by the following formula: $\sigma_f = 3FL/2bh^2$, where σ_f is the flexural strength (MPa), F is the failure load (N), L is the span (mm), b is the width of the specimen (mm), and h is the height of the specimen (mm). The compressive strength was determined by $\sigma_c = 4F/\pi d^2$, where σ_c is the compressive strength (MPa), F is the critical load (N), and d is the diameter of the specimen (mm).

Scanning electron microscopy (SEM; Merlin, Zeiss Company, Germany) was adopted to observe the microstructures of the mullite fibre-reinforced geopolymer composites fabricated under various conditions and the fractal surfaces resulting from the three-point bending test.

To determine the crystalline phases of the specimens, X-ray diffraction (XRD) tests were carried out (D8 Advance, Bruker Optics, Germany) using a scanning range from 10° to 80°. X-ray fluorescence spectrometer (XRF; ARL PERFORM X; Thermo Fisher, China) was used to determine the chemical compositions of the powder sample. Thermogravimetry (TG; STA8000, PerkinElmer, USA) combined with Fourier transform infrared spectroscopy (FTIR; Spectrum 3, PerkinElmer, USA) was used to simultaneously obtain the thermodynamic information and the functional groups of the specimens. Nitrogen with a flow rate of 20 ml/min was injected during the heat process, and the temperature was increased to 1600°C at a heating rate of 10°C/min.

High-temperature treatment was carried out in a muffle furnace (KSL-1750X; Kejing, China), heated at 5°C/min and held at the target temperature for 1 h, then cooled to room temperature to obtain the treated sample. Rectangular samples with dimensions of about 30 mm × 15 mm × 15 mm were used for high-temperature shrinkage test and microstructural analysis. Cutting and polishing of samples were carried out with a metallographic sample cutting machine (SYJ-50; Kejing, China) and a polishing machine (UNIPLO-802; Kejing, China).

2.3 RESULTS AND DISCUSSIONS

2.3.1 Interfaces between Geopolymer and Fibre

In fact, mullite is also one of the phases of geopolymer produced after calcination above 1150°C [21], but the proportion of mullite produced by polymerization is small. On the one hand, the addition of mullite fibre increases the weight proportion of the highest melting point component in the composite; furthermore, the fibrous structure can hinder the shrinkage of the matrix at high temperatures. The dispersion of the fibres is critical to the properties of composites. A typical SEM image of the dispersion of fibres in the slurry is shown in Figure 2.3, indicating excellent dispersion of mullite fibres using the processing described in Section 2.1.

Fractal surfaces after three-point bending tests were also examined using SEM. It was first found that geopolymer particles are grown on the surface of the mullite fibres (cf Figure 2.4a), which proved that mullite fibre has good compatibility with

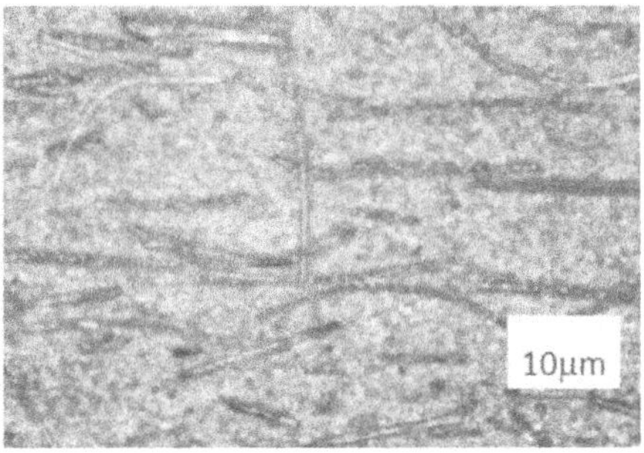

FIGURE 2.3 The fibres are evenly dispersed in the slurry [23].

the geopolymer matrix. Because the stiffness of the fibre is higher than that of the matrix, it is expected that the load will transfer from the geopolymer matrix to the fibres under loading, hence improving the strength of the matrix.

The fracture mechanism of the sample can also be inferred from the fractal surfaces of the samples. It is found that sections of fibres are detached from the matrix upon fracture (cf Figure 2.4b), which is referred to as fibre pull-out. It is well known that the additional energy consumed in the process of fibre pull-out improves the toughness of the material. These observations indicate that the bonding between the fibre surface and the matrix is desirable for both fibre pull-out and load transfer. The strengthening effect of fibres is also confirmed by observation during the bending test. The sample with no fibres split into two parts after fracture, exhibiting brittle fracture characteristics, while the sample with fibres remains attached after the test.

In some cracks of the sample, it can also be seen that the cracks are usually accompanied by crack deflection due to fibres, as indicated by arrows in Figure 2.4c. This shows that the fibre/matrix interface is weaker than the matrix. When the crack propagates, the crack tends to propagate along the weak fibre/matrix interface. Crack deflection can also lead to crack branching, as indicated by arrows in Figure 2.4c, which further increases the resistance to crack propagation.

2.3.2 Mechanical Properties

Figure 2.5 shows the mechanical properties of the samples post 7-day or 28-day curing. The maximum compressive strength reaches 27 MPa, and the maximum

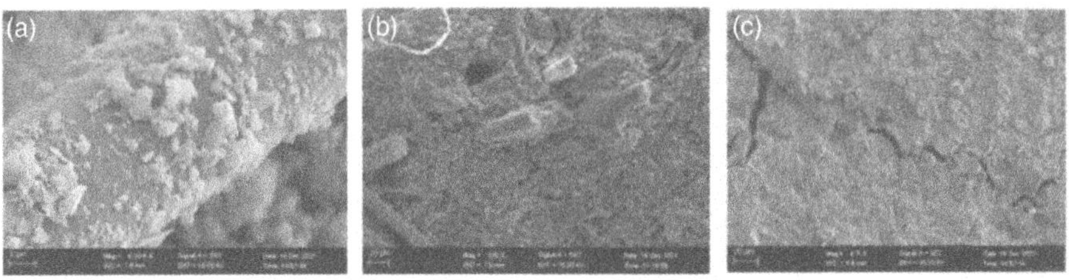

FIGURE 2.4 SEM photos of the interface between mullite fibre and geopolymer matrix, (a) geopolymer grains on mullite fibres, (b) fibre pull-out, and (c) crack deflection [23].

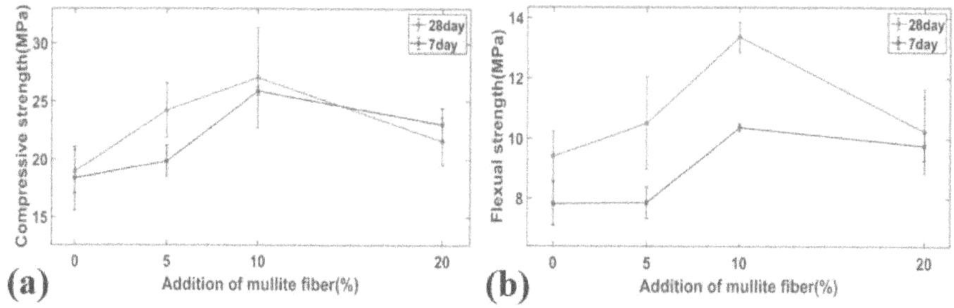

FIGURE 2.5 (a) Flexural and (b) compressive strength of geopolymer ceramic composites [23].

flexural strength reaches 13 MPa, both of which are achieved with the addition of 10% fibres. Compared with the sample without fibre addition, which has a compressive strength of 18.9 MPa and a flexural strength of 9.4 MPa, they are increased by more than 30%. With the increase in fibre addition, the overall mechanical properties first rise and then decrease, indicating that the addition of 20% fibres has caused some defects in the material. The mechanical properties of the samples cured for 28 days are slightly higher than those of the samples cured for 7 days. The data are provided in Table 2.3.

2.3.3 Phase Evolution at Elevated Temperatures

2.3.3.1 XRD Results

Figure 2.6 shows the XRD patterns of geopolymers after various heat treatments. Pattern (a) of the sample without any treatment after curing for 7 days reveals that there are quartz crystalline phase q and amorphous phase. After different heat treatments, the position of the diffraction peak does not change, but the intensity increases. These crystalline phases were identified as quartz and berlinite (aluminium phosphate) phases. After firing at 1150°C, peaks correlated with cristobalite, mullite, and aluminium phosphate phases appeared. With exposure to 1550°C, the intensity at about 21° of the crystal face (101) of cristobalite reaches a maximum. According to previous studies [27], the geopolymer with a complete reaction will

TABLE 2.3
The Effect of Fibre Content on Flexural and Compressive Strength of Geopolymer Composites

Addition of mullite fibre (%)	0	5	10	20
7-day flexural strength (MPa)	7.83 ± 0.72	7.86 ± 0.52	10.37 ± 0.12	9.76 ± 0.47
7-day compressive strength (MPa)	18.33 ± 2.76	19.83 ± 1.37	25.90 ± 0.37	23.03 ± 1.40
28-day flexural strength (MPa)	9.41 ± 0.83	10.51 ± 1.53	13.37 ± 0.50	10.25 ± 1.40
28-day compressive strength (MPa)	18.95 ± 1.89	24.23 ± 2.34	27.07 ± 4.32	21.60 ± 2.05

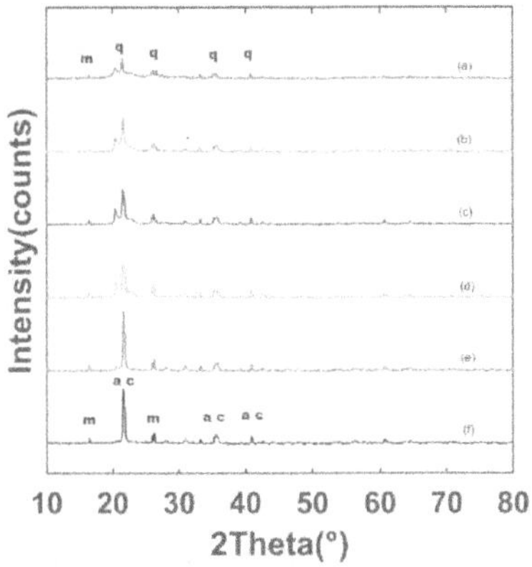

FIGURE 2.6 XRD patterns of the geoploymer sample with 20% fibre addition under various heat treatments. (a) Room temperature. (b) 1150°C calcination for 1 h. (c) 1250°C calcination for 1 h. (d) 1350°C calcination for 1 h. (e) 1450°C calcination for 1 h. (f) 1550°C calcination for 1 h. Peak symbols correspond to phases: m – mullite, c – cristobalite, a – aluminium phosphate, and q – quartz [23].

show an amorphous phase in XRD patterns. The remaining quartz peaks (~20°) in Figure 2.6a indicate that excessive quartz from metakaolinite is not fully consumed by the reaction.

2.3.3.2 FTIR Spectra

Figure 2.7 shows the FTIR spectra of different geopolymer samples. The addition of fibre had no significant effect on the curve. There are some obvious absorption peaks: 1083 cm^{-1} of Si-O-Si asymmetric stretching, 798 cm^{-1} of Al-O-Si, and 464 cm^{-1} of Si-O bending. The broad absorption around 950–1300 cm^{-1} corresponds to the stretching vibration of P-O-Si-O, and the band around 1600–1740 cm^{-1} is generated from the adsorbed water and P-O group of stretching and bending vibration. The broad absorption band around 3400–3700 cm^{-1} corresponds to the vibration of hydroxyl (OH) groups and H_2O molecules.

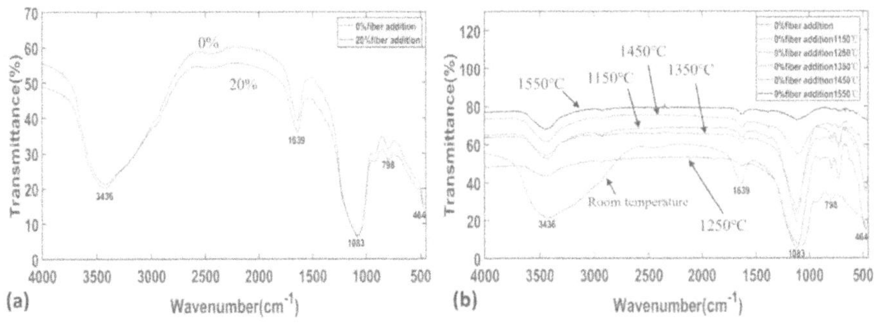

FIGURE 2.7 Infrared spectra of geopolymers treated with different fibre ratios and temperatures [23].

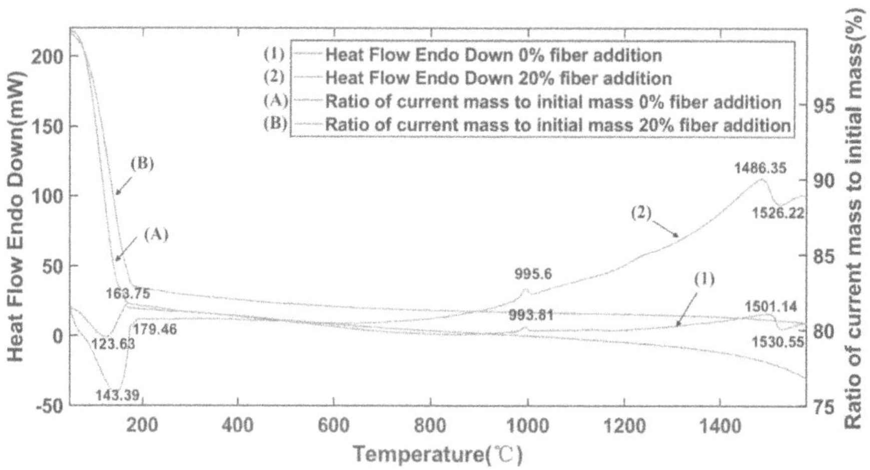

FIGURE 2.8 Results of differential thermal analysis of samples [23].

After calcination at high temperatures, obvious curve changes are found. First, it is noted that the wide absorption peak of 3436 cm^{-1} reduces, symbolizing the reduction of hydroxyl groups. Some changes have also taken place where the wavelength is less than 700 cm^{-1}. Based on the experimental results and previous studies [28, 29], the bonding reaction among the P-O tetrahedral units of the phosphoric acid solution, the Si-O tetrahedral units, and the Al-O layer of metakaolinite generated the polymeric structure of P-O-Si-O-Al-O, which provides the main structure of the acid-based geopolymer.

2.3.3.3 TG-DSC Curves

Thermogravimetry and differential scanning calorimeter (TG-DSC) curves of zero fibre addition and 20% fibre addition samples from 50°C to 1600°C were obtained and shown in Figure 2.8. Since the sample was not dried in advance, the weight loss of 17.27% was detected below 200°C of 20% fibre addition samples, mainly due to the loss of absorbed water and condensation of hydroxyl groups [30]. There is an obvious endothermic peak detected by the DSC curve at around 140°C. Continued mass loss of approximately 0.24% was observed between 200°C and 250°C because of dehydroxylation of the high-temperature hydroxyl group. A small exothermic peak was observed around 1000°C. The exothermic phenomenon was detected at 1400–1500°C, which was due to the formation of the mullite phase. An obvious endothermic peak was detected near 1530°C. Recalling the discussion about the disappearance of the $AlPO_4$ phase at 1550°C in Section 2.3.1, it is believed that this endothermic peak is related to the melting or decomposition of the $AlPO_4$ phase. After the temperature exceeds 1400°C, the weight loss of the sample becomes faster. Again, this is mainly attributed to the decomposition of $AlPO_4$, which coincides with the DSC analysis. It was found that the samples added with fibres reduced the weight

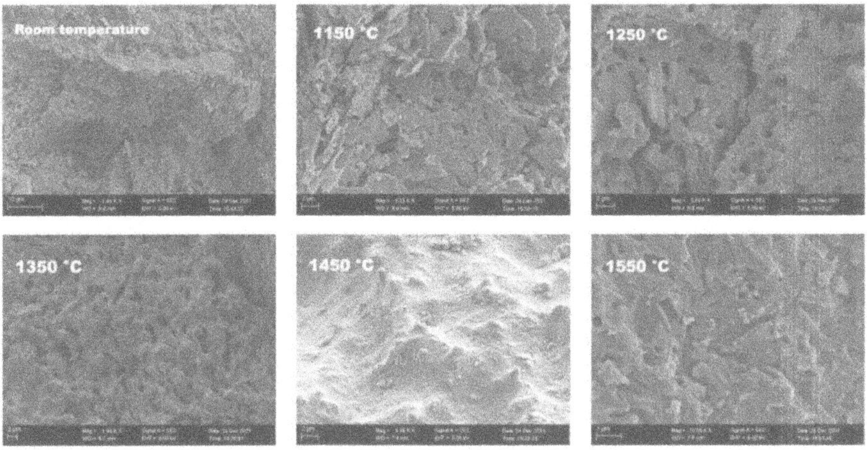

FIGURE 2.9 SEM photos of samples treated at different temperatures, the heat treatment indicated on the images [23].

loss by about 4% compared to monolithic geopolymer up to 1600°C according to the TG curve, owing to the proportion of stable mullite fibres.

2.3.3.4 SEM Observation

As shown in Figure 2.9, the geopolymer at room temperature is mainly composed of lamellar structure. At 1250°C, the lamellar structure starts to produce some connections. When the temperature rises to 1350°C, the lamellar structure has formed a porous structure. After 1450°C treatment, some fine needle-like structures are produced on the sample surface, speculatively mullite and cristobalite. When the temperature reaches 1550°C, the needle-like structure has evolved into a columnar structure. It is believed that the columnar crystal is mainly due to mullite and cristobalite crystal growth in the liquid phase formed by the melting of aluminium phosphate above 1500°C.

2.3.4 Thermal Properties

After the high-temperature treatment of the samples, linear shrinkage was obtained, as shown in Figure 2.10. The data are provided in Table 2.4. Each data point is the average of three samples. The linear shrinkage of the sample increased steadily and slightly until 1350°C, reaching about 3–5%, but suddenly the shrinkage at 1450°C exceeded 10%. Generally, the linear shrinkage of the sample with 20% fibre is about 30% lower than that of the sample without fibre addition.

Below 1350°C, the sample can basically maintain the shape, and some microcracks can be observed on the surface. Above 1450°C, the sample has undergone serious deformation and cracks. This coincides with the weight loss and exothermic phenomenon that occurred in TG-DSC above 1400°C. Above 1550°C, the deformation

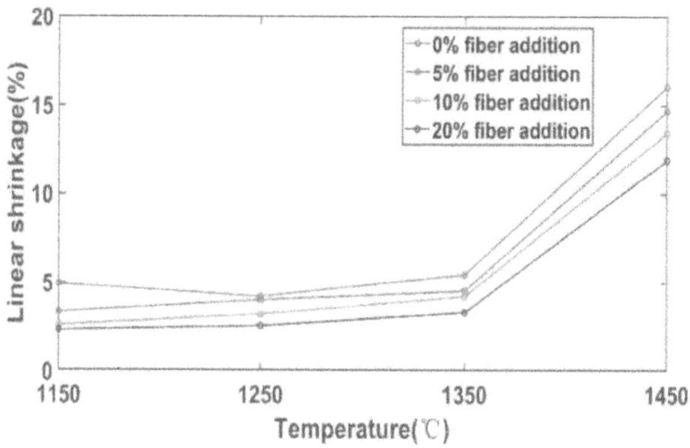

FIGURE 2.10 Linear shrinkage at different temperatures [23].

TABLE 2.4
Effects of Fibre Content on the Linear Shrinkage of Geopolymer Composites after Calcination

Addition of Mullite Fibre (%)	0	5	10	20
Shrinkage at 1150°C	4.92%	3.35%	2.61%	2.33%
Shrinkage at 1250°C	4.22%	4.02%	3.21%	2.56%
Shrinkage at 1350°C	5.44%	4.53%	4.23%	3.32%
Shrinkage at 1450°C	16.05%	14.68%	13.42%	11.90%

becomes more serious and even shows the trend of flow. This is correlated with the TG-DSC analyses described in Section 2.3.3.

It is noted that not only the linear shrinkage of the sample decreases with the addition of fibres but also the number of cracks on the sample surface decrease and are invisible with the addition of more than 10% fibres, as shown in Figures 2.11 and 2.12. The results confirm that the addition of mullite fibres has hindered the shrinkage and cracking effect during heat treatment.

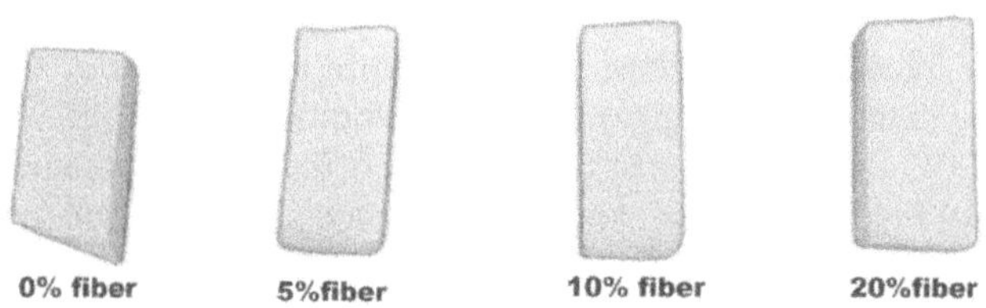

FIGURE 2.11 The surface crack of the sample after 1150°C treatment for 1 h [23].

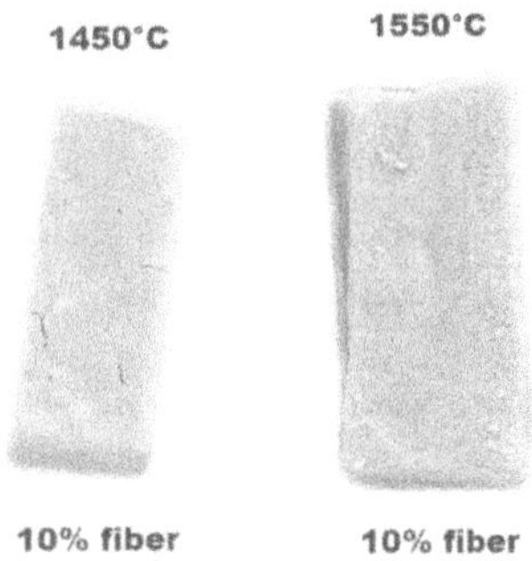

FIGURE 2.12 The surface of 10% fibre added samples after 1450°C and 1550°C treatment [23].

2.4 CONCLUSIONS

This chapter investigated several important properties, such as thermal resistance, mechanical properties, and interfaces of geopolymer ceramics reinforced with mullite fibres. These particular fibre-reinforced geopolymer composites were prepared from kaolin and mullite fibres with phosphoric acid as an activator. XRD, TG-DSC, FTIR, and SEM were used to determine the phase evolution and strengthening mechanisms. With the addition of mullite fibres, the mechanical properties increased by at least 20%. The optimum flexural strength exceeded 13 MPa. It was found that mullite fibres have desirable interface bonding with this type of geopolymer, promoting both crack deflection and fibre pull-out strengthening mechanisms. This is correlated with the significant strengthening effect of the fibres. The linear shrinkage when treated at 1150–1450°C was investigated and correlated with XRD analyses. The addition of mullite fibres reduced the linear shrinkage significantly up to 1350°C. The large linear shrinkage at 1450°C was correlated with the melting and decomposition of the $AlPO_4$ phase. In summary, the following conclusions can be drawn:

1. Mullite fibre has a good affinity for phosphate geopolymer and has a significant fibre reinforcement effect. The addition of 10% mullite fibre is optimum for the mechanical properties, reaching a 28-day compressive strength over 27 MPa and flexural strength over 13 MPa, while the addition of 20% has the strongest inhibitory effect on the shrinkage after high-temperature calcination, which has the linear shrinkage of 3.32% after 1350°C calcination for 1 h, 30% lower than that without fibre addition. The main strengthening mechanisms of the fibre are fibre pull-out and fibre deflection effects.
2. Although a certain amount of microcracks appeared in the geopolymer at 1350°C for 1 h, it can still maintain its shape, and the inhibition effect on the crack is more obvious with the increase in fibre addition. When it reaches above 1450°C, there obvious deformation, with 11.9% linear shrinkage of 20% fibre addition sample. Signs of melting appear at 1550°C.
3. The polymeric structure of P-O-Si-O-Al-O is the main structure of geopolymer activated by phosphoric acid. When heated above 1150°C, its main

forms are mullite and calcite. When heated to 1450°C and above, $AlPO_4$ phase is formed, which is responsible for the melting of the composite above 1550°C. This limits the thermal resistance of the composites.

To further improve the thermal and mechanical properties of phosphoric acid-activated geopolymer composites will require a more efficient dispersion process for higher fibre content, e.g. 20%, and/or optimization of the composition to hinder the formation of the $AlPO_4$ phase.

ACKNOWLEDGEMENTS

The financial support of Tsinghua-Foshan Advanced Manufacturing Institute is acknowledged. The authors are grateful for the support of colleagues at the Future Materials & Design Research Center, The Future Lab, Tsinghua University. The assistance of the technical staff at the School of Materials Science and Engineering Analysis Center, Tsinghua University is also acknowledged.

REFERENCES

1. Davidovits J. Geopolymers. *Journal of Thermal Analysis* 1991, **37**: 1633–1656.
2. Davidovics M, Davidovits J, Orlinski J, *et al.* Geopolymers form room temperature ceramic matrix. *Materials and Processing Report* 1988, **3**: 5.
3. Lahoti M, Tan KH, Yang EH. A critical review of geopolymer properties for structural fire-resistance applications. *Construction and Building Materials* 2019, **221**: 514–526.
4. Zhang Z, Provis JL., Reid A, *et al.* Geopolymer foam concrete: An emerging material for sustainable construction. *Construction and Building Materials* 2014, **56**: 113–127.
5. Hamdi N, Ben MI, Srasra E, *et al.* Production of geopolymer binders using clay minerals and industrial wastes. *Comptes Rendus Chimie* 2019, **22**: 220–226.
6. Steinerova M, Matulova L, Vermach P, *et al.* The brittleness and chemical stability of optimized geopolymer composites. *Materials (Basel)* 2017, **10**.
7. Sun ZQ, Cui H, An H, *et al.* Synthesis and thermal behavior of geopolymer-type material from waste ceramic. *Construction and Building Materials* 2013, **49**: 281–287.
8. Morsy MS, Rashad AM, Shoukry H. Potential use of limestone in metakaolin-based geopolymer activated with H_3PO_4 for thermal insulation. *Construction and Building Materials* 2019, **229**: 117088.
9. Alomayri T, Vickers L, Shaikh FUA, *et al.* Mechanical properties of cotton fabric reinforced geopolymer composites at 200–1000°C. *Journal of Advanced Ceramics* 2014, **3**: 184–193.
10. Furtos G, Silaghi-Dumitrescu L, Pascuta P, *et al.* Mechanical properties of wood fibre reinforced geopolymer composites with sand addition. *Journal of Natural Fibres* 2019, **18**: 285–296.
11. Malenab RAJ, Ngo JPS, Promentilla MAB. Chemical treatment of waste abaca for natural fibre-reinforced geopolymer composite. *Materials (Basel)* 2017, **10**.
12. Korniejenko K, Frączek E, Pytlak E, *et al.* Mechanical properties of geopolymer composites reinforced with natural fibres. *Procedia Engineering* 2016, **151**: 388–393.
13. Liew YM, Kamarudin H, Mustafa ABAM, *et al.* Processing and characterization of calcined kaolin cement powder. *Construction and Building Materials* 2012, **30**: 794–802.
14. Du FP, Xie SS, Zhang F, *et al.* Microstructure and compressive properties of silicon carbide reinforced geopolymer. *Composites Part B: Engineering* 2016, **105**: 93–100.

15. Rahman AS, Radford DW. Cure cycle optimization of an inorganic polymer matrix material for high temperature fibre reinforced composites. *Composites Part A: Applied Science and Manufacturing* 2016, **85**: 84–93.
16. Fang Y, Wang A, He K, *et al.* Property evolution of geopolymer composites with SiC whiskers loaded with BN coating at elevated temperatures. *Construction and Building Materials* 2021, **309**.
17. Yan S, He PG, Zhang Y, *et al.* Preparation and in-situ high-temperature mechanical properties of C_f-SiC_f reinforced geopolymer composites. *Ceramics International* 2017, **43**: 549–555.
18. Liu LP, Cui XM, Qiu SH, *et al.* Preparation of phosphoric acid-based porous geopolymers. *Applied Clay Science* 2010, **50**: 600–603.
19. Sellami M, Barre M, Toumi M. Synthesis, thermal properties and electrical conductivity of phosphoric acid-based geopolymer with metakaolin. *Applied Clay Science* 2019, **180**: 105192.
20. Cao D, Su D, Lu B, *et al.* Synthesis and structure characterization of geopolymeric material based on metakaolininte and phosphoric acid. Journal of the Chinese Ceramic Society 2005, **33**: 1385–1389.
21. Liu L, Cui X, He Y, *et al.* The phase evolution of phosphoric acid-based geopolymers at elevated temperatures. *Materials Letters* 2012, **66**: 10–12.
22. Reinders L, Pfeifer S, Kröner S, *et al.* Development of mullite fibres and novel zirconia-toughened mullite fibres for high temperature applications. *Journal of the European Ceramic Society* 2021, **41**: 3570–3580.
23. Wei Q, Liu Y, Le H. Mechanical properties and thermal stability of mullite fibre reinforced phosphoric acid activated metakaolinite geopolymer composites. *Materials* 2022, **15**(12): 4185.
24. Prashanth S, Subbaya KM, Nithin K, *et al.* Fibre reinforced composites – A review. *Journal of Material Science and Engineering* 2017, **6**: 2–6.
25. Wang H, Li HH, Yan FY. Synthesis and mechanical properties of metakaolinite-based geopolymer. *Colloids and Surfaces A: Physicochemical and Engineering Aspects* 2005, **268**: 1–6.
26. Wan Q, Rao F, Song SX, *et al.* Reexamining calcination of kaolinite for the synthesis of metakaolin geopolymers – roles of dehydroxylation and recrystallization. *Journal of Non-Crystalline Solids* 2017, **460**: 74–80.
27. Lassinantti GM, Romagnoli M, Pollastri S, *et al.* Inorganic polymers from laterite using activation with phosphoric acid and alkaline sodium silicate solution: Mechanical and microstructural properties. *Cement and Concrete Research* 2015, **67**: 259–270.
28. Gao L, Zheng Y, Tang Y, *et al.* Effect of phosphoric acid content on the microstructure and compressive strength of phosphoric acid-based metakaolin geopolymers. *Heliyon* 2020, **6**: e03853.
29. Prud'homme E, Michaud P, Joussein E, *et al.* In situ inorganic foams prepared from various clays at low temperature. *Applied Clay Science* 2011, **51**: 15–22.
30. Wang YS, Dai JG, Ding Z, *et al.* Phosphate-based geopolymer: Formation mechanism and thermal stability. *Materials Letters* 2017, **190**: 209–212.

3 Development of Geopolymers Using Industrial Wastes

Sirithan Jiemsirilers

3.1 INTRODUCTION

Alternative cementitious materials called "geopolymers" are gaining increasing attention because of their low carbon emission and energy consumption. It is synthesised by SiO_2 and Al_2O_3 in an alkali activator solution at the temperature range of 20–120°C [1]. Geopolymers utilise industrial wastes and by-products, such as fly ash, blast furnace slag, rice husk ash, bottom ash, aluminium waste, and mining tailings waste. As a result, these wastes can also be used as starting materials for geopolymers. Mechanical properties of geopolymers rely on large aluminosilicate chains in three-dimensional structures. Therefore, the source of Al and Si becomes important, especially in cases of waste reuse.

The primary objective of early geopolymer synthesis was to develop a material to replace cement, as it is well known that the production of Portland cement requires high heat energy, ranging from 1400 to 1600°C. In this process, raw materials containing alumina (Al_2O_3), silica (SiO_2), and calcium oxide (CaO) are burned to produce Portland cement. Additionally, the cement manufacturing process releases significant amounts of greenhouse gases into the atmosphere, accounting for up to 7% of the total global emissions, contributing greatly to global warming.

Therefore, geopolymers are a new type of binder that has been developed with the potential to replace traditional cement and concrete materials in the future. In recent years, geopolymer technology has been applied in various industries, including construction, decorative arts, and even the electrical and aerospace industries. Today, this technology is gaining popularity internationally.

Geopolymers synthesised from calcined kaolin have been extensively researched, with numerous studies published on factors affecting geopolymer properties, such as raw material ratios, water content, and types of alkalis used. Therefore, this study focused on synthesising geopolymers using kaolin residue with low alkaline concentration. The modification of the forming process and the emphasis on developing materials with sufficient strength to create products align with both environmental sustainability and waste management. This approach can provide a practical and eco-friendly solution to managing industrial waste and by-products.

DOI: 10.1201/9781003535683-3

3.2 MATERIALS AND EXPERIMENTS

3.2.1 Raw Materials and Chemicals

Raw materials and chemicals used in this study are as follows. Metakaolin-based geopolymer was prepared using metakaolin, kaolin residue, and calcined kaolin residue.

- Kaolin residue, Imerys Ceramics (Thailand) Company Limited
- Metakaolin, Hayashi Kasei Company Limited, Japan
- Sodium hydroxide flakes (98% purity), PT Interchem Plasagro JAYA, Indonesia
- Sodium hydroxide (97% purity) pellets, Nacalai Tesque Inc., Japan
- Potassium hydroxide (97% purity) pellets, Nacalai Tesque Inc., Japan
- Sodium silicate solution (Na_2SiO_3), C. Thai Chemical Company Limited, Thailand
- Na_2SiO_3 solution, Japan
- Distilled water

3.2.2 Preparation of Raw Materials

Prior to the geopolymerisation process, calcined kaolin residue was obtained by calcination of kaolin residue at 700°C in air for 1 h with a heating rate of 2°C/min by an electric furnace, HP Advanced Ceramic Company Limited, Thailand. The alkali solution, NaOH solutions were prepared in the concentrations of 4, 6, 8, and 10 M for synthesis of calcined kaolin residue-based geopolymer. Moreover, NaOH and KOH with concentrations of 0.5, 1, 3, 5, and 10 M were obtained by dissolving dried pellets of NaOH and KOH and then used to synthesise metakaolin-based geopolymer.

The alkali activator solution was a mixture of Na_2SiO_3 solution and NaOH solution, as well as Na_2SiO_3 solution and KOH solution. The Na_2SiO_3/NaOH ratios of 0.44–3.43 (by weight) were used to prepare the geopolymer produced from calcined kaolin residue. Moreover, the Na_2SiO_3/NaOH and Na_2SiO_3/KOH ratios of 2.25 (by weight) were employed to fabricate the geopolymer produced from metakaolin.

3.2.3 Characterisation of Raw Materials

3.2.3.1 Characterisation of Kaolin Residue and Calcined Waste

The chemical compositions of kaolin residue and calcined kaolin residue were analysed by using X-ray fluorescence (XRF), S8 Tiger, Bruker. X-ray diffraction (XRD) (X'Port Pro, PANalytical) was used to determine the phase component of kaolin residue and calcined waste. XRD diffractograms of these wastes were recorded in the 2θ range of 5°–80° at 0.02° step size. Field emission scanning electron microscope (FESEM) JEOL 7800F and energy dispersive X-ray spectroscopy (EDX) were employed to observe the morphology and element composition, respectively, of kaolin residue. The particle size of kaolin residue was examined using Mastersizer 2000, Malvern Instruments.

3.2.3.2 Characterisation of Metakaolin

The chemical component of metakaolin powders was measured via XRF, ZSX Primus II, Rigaku. The morphology of metakaolin was examined via XRD analysis using Smart Lab 3kw, Rigaku. A scanning electron microscope (SEM) (TM303Plus, Hitachi) was used to observe the plate-like structure of metakaolin. The particle size of metakaolin was determined via a laser diffraction particle size analyser, SALD-7000, Shimadzu.

3.3 PREPARATION OF GEOPOLYMERS

Kaolin residue or waste from the kaolin washing industry is generated daily at approximately 100,000 tons/year [2]. Improper disposal of this waste has negative effects on both the environment and society, causing widespread concern among academics around the world. Given the high proportion of SiO_2 and Al_2O_3 contents in kaolin residue, it can be used as precursor materials to form geopolymers for building materials. Reducing waste is recognised as an effective and sustainable strategy.

Geopolymers are generally produced by the casting method with a high content of alkali activator solution. This ensures that the geopolymer paste is usable [2, 3]. Moreover, it is well known that the production process of alkali activator solution causes higher energy consumption and carbon dioxide emissions. Highly alkaline activator solutions can be problematic in terms of sustainability and environmental friendliness. The use of a large amount of alkali activator solution also encourages the formation of pores. This information results in a subsequent decrease in the mechanical performance and durability of the geopolymers. Previous investigations have provided evidence that 28-day compressive strength achievement for cast geopolymers is limited under room temperature [4–8]. This problem is solved by using geopolymer synthesis with a low amount of alkali activator solution and the preparation of the geopolymer sample using a pressing method. Previous research has indicated that the improved compressive strength and low porosity of geopolymer were prepared by pressing approach [9, 10]. These properties ensure that the pressed geopolymers can be used in applications such as bricks and paving blocks.

The geopolymer samples were prepared using pressing and casting methods, which have lower and higher alkali activator solution contents, respectively, as shown in Figures 3.1 and 3.2. To prepare the geopolymer using the pressing method,

FIGURE 3.1 Calcined kaolin residue-based geopolymer prepared by pressing method with a height of 50 mm and a diameter of 25 mm.

FIGURE 3.2 Calcined kaolin residue-based geopolymer prepared with casting method with a width of 50 mm and length of 50 mm.

homogeneous semidry powders were produced by mixing calcined kaolin residue with a lower alkali activator solution in a Hobart mixer for 5 minutes; the alkali activator solution concentration is between 20–26% by weight. The cylindrical stainless-steel mould (a diameter of 25 mm) containing these semidry powders was compressed by using the hydraulic press machine at 20 MPa pressure to obtain a specimen with a height of 50 mm.

For the preparation of the geopolymer using the casting method, calcined kaolin residue was mixed with the higher alkali activator solution for 5 min in a Hobart mixer. In this case, the alkali activator solution comprises between 46% and 52% of alkali by weight. The geopolymer pastes were drained into a silicone mould with dimensions of 50 mm × 50 mm × 50 mm, and it was oscillated on the vibrating table for 5 min. After 24 h, the geopolymer samples were removed from the mould, as shown in Figure 3.2. Following the demoulding of the geopolymer samples, they were covered with plastic film to prevent water evaporation and kept at room temperature for 7 days ±3 h, according to ASTM C109/C109M-02 [11].

3.4 RESULTS AND DISCUSSIONS

3.4.1 Chemical Composition of Kaolin Residue and Calcined Kaolin Residue

The results of the chemical composition of kaolin residue and calcined kaolin residue analysed by XRF are illustrated in Table 3.1. Kaolin residue consists of a high proportion of SiO_2 and Al_2O_3, which are equal to 51.21 and 30.98 wt%, respectively. Calcined kaolin residue has a high proportion of SiO_2 and Al_2O_3, which are equal to 55.56 and 33.85 wt%, respectively. The amounts of SiO_2 and Al_2O_3 of calcined kaolin residue were higher than those of kaolin residue waste. Moreover, the amounts

TABLE 3.1
Chemical Compositions of Kaolin Residue and Calcined Kaolin Residue [13]

Composition	Kaolin Residue Waste (wt%)	Calcined Kaolin Residue Waste (wt%)	Metakaolin [12] (wt%)
SiO_2	51.21	56.56	53.94
Al_2O_3	30.98	33.85	44.43
Na_2O	0.02	0.23	–
K_2O	5.20	5.15	0.87
CaO	0.02	0.05	0.04
MgO	1.19	0.05	0.08
TiO_2	0.16	0.16	–
Fe_2O_3	2.22	2.16	0.63
MnO	0.15	0.14	–
SO_3	–	–	0.01
Loss on ignition (LOI)	8.85	1.65	–
Total	100.00	100.00	100.00

of SiO_2 and Al_2O_3 of calcined kaolin residue were higher and lower than those of metakaolin in the research of Tippayasam et al. [12], respectively.

3.4.2 Microstructure of Raw Materials

Figure 3.3 shows the SEM micrograph of kaolin residue. The morphology of the kaolin residue particles presented hexagonal, rod, corroded, plate, sharp edge, and irregular shape crystals as indicated in Figure 3.3a. Kaolinite and halloysite crystals

FIGURE 3.3 Morphology of kaolin residue (a) and kaolinite (Point a1) and halloysite (Point a2) in kaolin residue [12].

in the kaolin residue remaining from the kaolin washing process are Point a1 and Point a2, respectively. These crystals corresponded to the crystalline phases of well-ordered kaolinite, halloysite, muscovite, mica, sanidine, and quartz, respectively, and these results were strongly supported by X-ray diffractogram.

3.4.3 XRD Analysis

From XRD analysis as shown in Figure 3.4, the phase composition of kaolin residue was halloysite ($Al_2O_3{\cdot}2SiO_2{\cdot}2H_2O$), kaolinite ($Al_2O_3{\cdot}2SiO_2{\cdot}2H_2O$), albite ($NaAlSi_3O_8$), sanidine (K, Na)($AlSi_3O_8$), muscovite ($KAl_2(AlSi_3O_{10})(F, OH)_2$), and quartz ($SiO_2$). Geopolymerisation reaction was conducted for calcined kaolin residue when it was mixed with a low amount of alkali activator solution (24 wt%) by Hobart mixer. The homogeneous mixture was poured into a mould with a size of ø 25 × 50 mm and pressed by using a hydraulic pressing machine at 21 MPa. The geopolymer sample was removed and covered with plastic film to avoid water evaporation and cured at room temperature for 7 days ±3 h before characterisation in accordance with ASTM C109 [11].

3.4.4 Microstructure of Geopolymers Produced from Different Forming Methods

For the pressing method in Figure 3.5, the semidry powder that resulted from mixing the calcined waste with the lower content of the alkali activator solution was compressed using a hydraulic press machine. After completing the mixing, the semidry powders were poured into a steel mould, and the pressures were applied in the mould to eliminate large

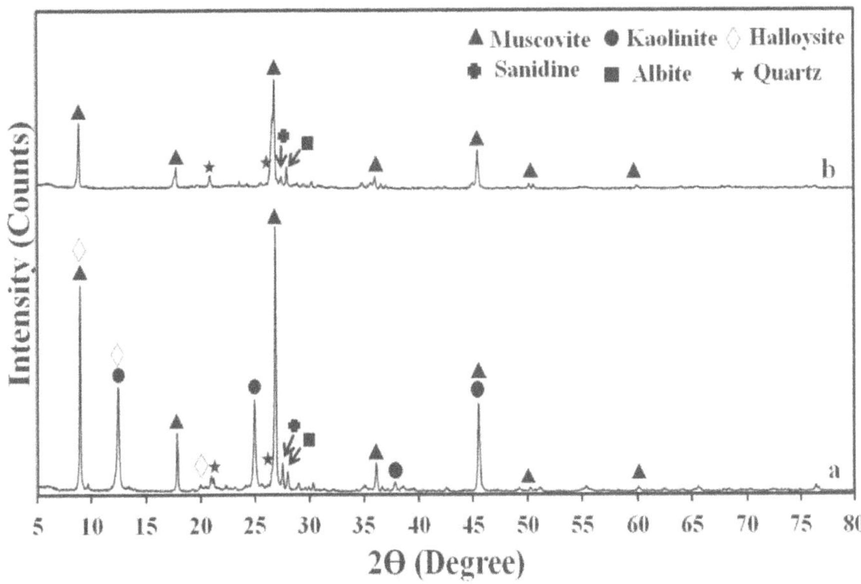

FIGURE 3.4 XRD diffractograms of (a) kaolin residue and (b) calcined kaolin residue at 700 °C [13].

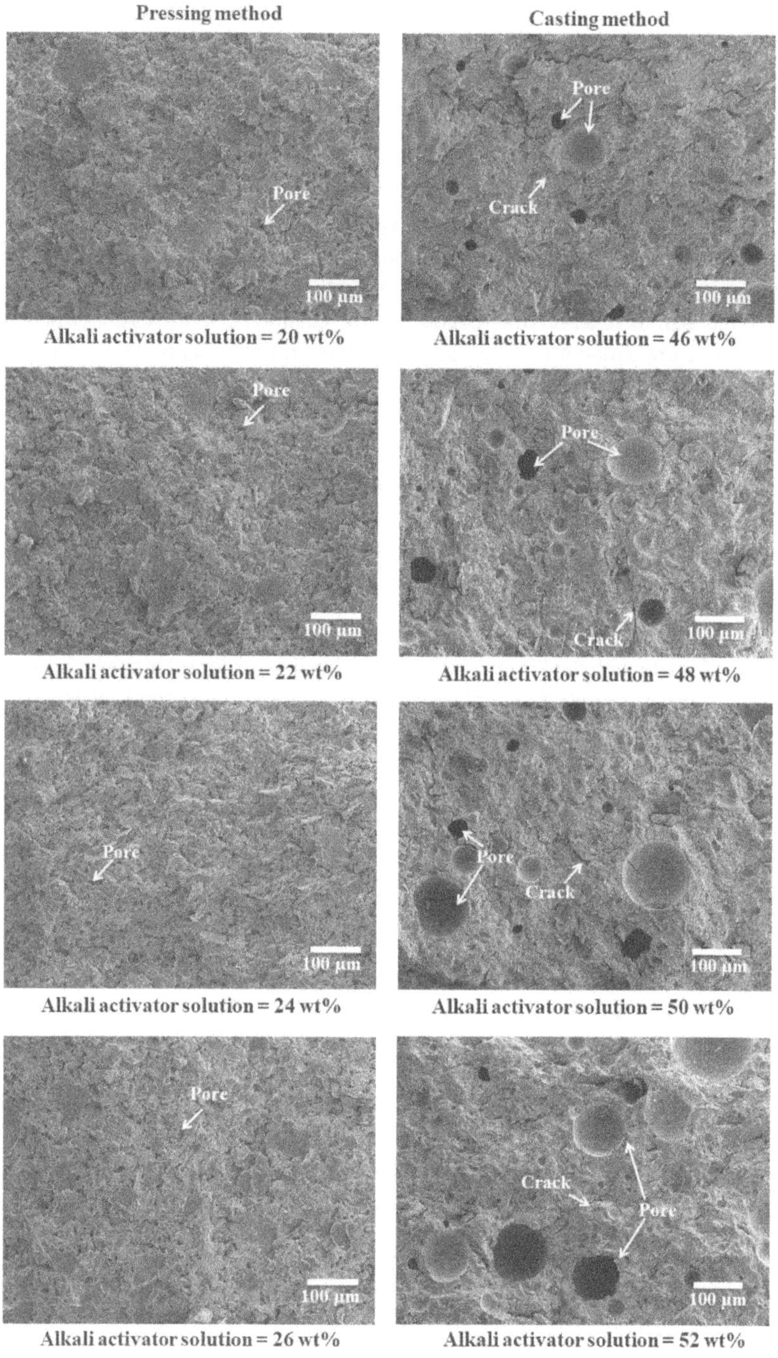

FIGURE 3.5 SEM micrographs of pressed and cast geopolymers with different contents of alkali activator solution.

air volumes (between the powder particles) to produce the compact matrix. The powders were highly compacted, and this led to the formation of a geopolymerisation reaction on the adjacent surface between the alkali activator solution and wet calcined waste particles. Thus, the surfaces of these particles in the compact metrics were dissolved by the alkali activator solution, thereby producing Si^{4+} and Al^{3+} ions in the solution. These ions were coordinated in silicate tetrahedral ($[SiO(OH)_3]^-$) and aluminate tetrahedral ($[Al(OH)_4]^-$) monomers in the alkali activator solution. Subsequently, the monomers were incorporated by polymerisation reaction into a stable oligomer which was the geopolymeric gel. The geopolymeric gel can fill into the pores of geopolymer specimens and lead to the formation of denser microstructures with reduced pore contents.

For the casting method in Figure 3.5, the geopolymer paste was obtained by mixing the calcined waste with the higher content of alkali activator solution. This paste was cast in a silicone mould and then vibrated on the vibrating table. A greater number of air bubbles moved toward the surface of the geopolymer paste as the content of the alkali activator solution was increased. After 1 day, the geopolymer samples were removed from their moulds after the initial hardening of geopolymer paste had occurred, and the samples were then allowed to cure for 7 days. After the geopolymerisation reaction, many air bubbles or larger pores remained in the geopolymer samples because the geopolymer paste had a high viscosity and a lower workability [14]. Moreover, this work proposes the reason for the cracking of geopolymer samples. It is suggested that the higher alkali activator solution contents used for the cast geopolymer resulted in higher shrinkage and the subsequent formation of cracks because the content of alkaline activator solution exceeds the requirement for the geopolymerisation process which may weaken the geopolymer structure. Moreover, the excess content of the alkali activator solution also hindered the process of geopolymerisation and the formation of the sodium carbonate phase, sodium hydroxide-like phase, and zeolite phase. Therefore, the larger pores, the higher crack density, and the hindering process of geopolymerisation combined were expected to adversely impact the compressive strength of these samples.

3.4.5 Mechanical Properties of Geopolymers

Figure 3.6 illustrates the influence of geopolymers containing low and high contents of alkali activator solution by pressing and casting on the compressive strength. For the pressed geopolymers, the compressive strength increased with the increment of the content of the alkali activator solution from 20 to 24 wt% but remained constant from 24 to 26 wt%. Therefore, the content of the alkali activator solution of 24 wt% was used to synthesise geopolymer in the next part. With respect to lower alkali activator solution proportions, the 20 and 22 wt% alkali activator samples had lower compressive strengths, i.e., 12.55 and 18.39 MPa, respectively. On the other hand, samples produced with higher proportions of alkali activator solution (i.e., 24 and 26 wt%) obtained higher compressive strengths (i.e., 26.98 and 27.74 MPa, respectively). The compressive strength of geopolymers produced with 24 and 26 wt% alkali activator solution increased approximately 114.98% and 121.04% compared to those samples produced with 20 wt% alkali activator solution. Therefore, lower proportions of alkali activator solution reduce the degree of the geopolymerisation reaction, which in turn reduces the compressive strength. Higher proportions of alkali activator solution

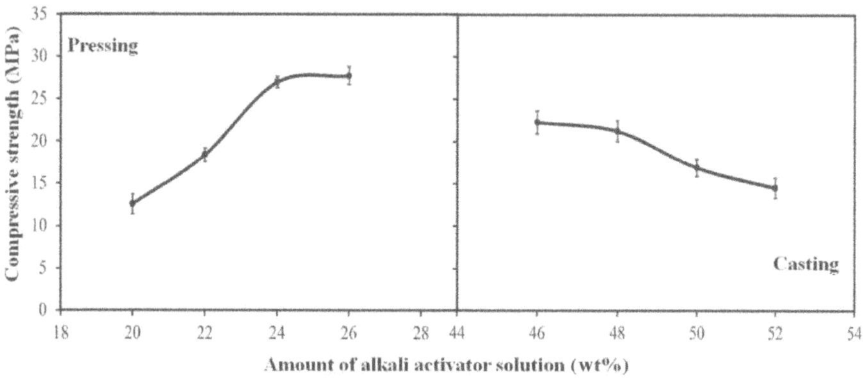

FIGURE 3.6 Compressive strength of pressed and cast geopolymers with different contents of alkali activator solution [20].

resulted in higher compressive strengths for the pressed geopolymers because of the higher dissolution capacities for the Al^{3+} and Si^{4+} ions from the calcined kaolin processing waste, which increased the formation of monomers and geopolymeric gel in the geopolymerisation process. Leong et al. [15] and Duxson et al. [16] reported that the amount of aluminosilicate gel formed was related to the compressive strength of geopolymers (i.e., increased gel formation leads to higher compressive strengths).

From SEM micrographs in Figure 3.7, the amount of geopolymeric gel increased with increasing content of alkali activator solution. This gel could fill in the pores of the pressed geopolymer samples, which resulted in denser samples and higher compressive strengths. Moreover, the lower water absorptions, the slightly increased bulk densities, and the lower apparent porosities also contributed to these results.

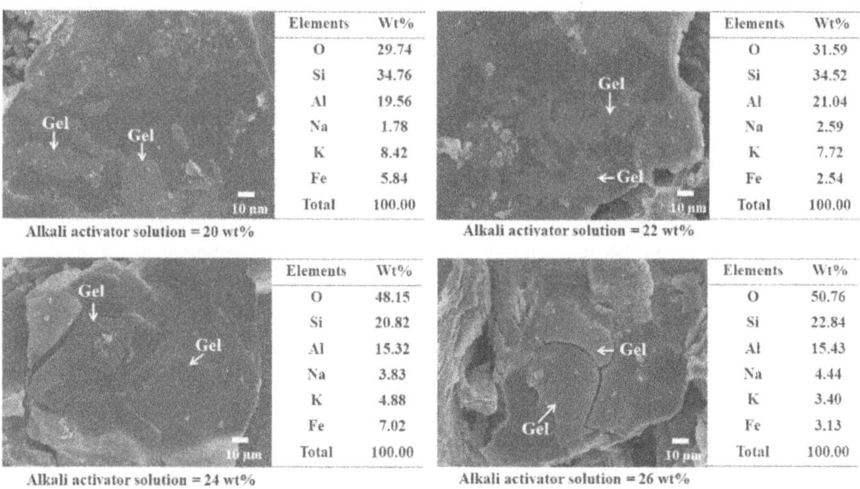

Alkali activator solution = 20 wt%

Elements	Wt%
O	29.74
Si	34.76
Al	19.56
Na	1.78
K	8.42
Fe	5.84
Total	100.00

Alkali activator solution = 22 wt%

Elements	Wt%
O	31.59
Si	34.52
Al	21.04
Na	2.59
K	7.72
Fe	2.54
Total	100.00

Alkali activator solution = 24 wt%

Elements	Wt%
O	48.15
Si	20.82
Al	15.32
Na	3.83
K	4.88
Fe	7.02
Total	100.00

Alkali activator solution = 26 wt%

Elements	Wt%
O	50.76
Si	22.84
Al	15.43
Na	4.44
K	3.40
Fe	3.13
Total	100.00

FIGURE 3.7 SEM micrographs of geopolymeric gel in pressed geopolymer with different contents of alkali activator solution.

For the cast geopolymers, the compressive strength clearly decreased when the content of the alkali activator solution was increased. The compressive strengths of cast geopolymers produced with alkali activator solutions of 46–52 wt% were 22.30–14.53 MPa, respectively. The lowest compressive strength of the cast geopolymers was measured for samples produced with 52 wt% alkali activator solution; the strength of these samples decreased by approximately 53.48% in comparison to samples produced with 46 wt% alkali activator solution. The geopolymers prepared with lower proportions of alkali activator solution exhibited higher compressive strengths than the geopolymers prepared with higher content of alkali activator solution. This can be explained by the pores in geopolymer samples. The geopolymer prepared with the higher content of alkali activator solution (52 wt%) had the highest apparent porosity and indicated the most porous microstructure and highest crack density per the SEM image in Figure 3.7. Furthermore, one factor affecting the strength of the cast geopolymers was the content of the alkali activator solution. The content of alkali activator solution within the samples increased the Si^{4+} and Na^{+} ions available for geopolymerisation, which resulted in a decrease in the compressive strength. The excessive Si^{4+} ion content inhibited geopolymerisation due to the precipitation of aluminosilicate [17]. The excessive Na^{+} ion content in the presence of CO_2 in air may result in the formation of Na_2CO_3. The presence of Na_2CO_3 may decrease the compressive strength of geopolymers [18] since Na_2CO_3 reduces the degree of geopolymerisation because of the high alkali concentration and due to the relatively weak binding and exchangeability of Na in the geopolymer structure [19]. FTIR spectra and SEM/EDX in this work support the formation of Na_2CO_3 in the cast geopolymer samples.

Additionally, the compressive strengths of pressed geopolymer samples were directly compared to cast geopolymer samples. The results indicate that processing has a very important effect on the compressive strength of geopolymers. The compressive strength of the pressed geopolymer was approximately 24.39% higher than that of the cast geopolymer. This result could be explained in many ways. First, the higher compressive strength could be attributed to sample compaction, fewer pores, and no observable cracks (per the SEM images of the pressed geopolymers). Furthermore, the lower water absorption, the higher bulk density, and the lower apparent porosity of the pressed geopolymers are also expected to enhance the compressive strength. This result is assumed due to the smaller proportion of alkali activator solution used in the pressed geopolymers compared to the cast geopolymers. In this study, it was clearly demonstrated that the pressing method resulted in higher compressive strengths than the casting method.

3.4.6 Mechanism of Geopolymer Formation of Pressed Geopolymer

The mechanism of geopolymer formation of pressed geopolymer with low alkali activator solution is presented in Figure 3.8. The compacted geopolymer matrix caused a high geopolymerisation reaction and low porosity in the samples because the surfaces of particles in the compacted geopolymer matrix were dissolved with an alkali activator solution to obtain the dissolved Si^{4+} and Al^{3+} ions. These ions resulted in the formation of aluminate tetrahedral monomers ($[Al(OH)_4]^-$) and silicate tetrahedral monomers ($[SiO(OH)_3]^-$) and the stable oligomers in an alkali activator solution. These oligomers were polycondensed as a three-dimensional aluminosilicate network.

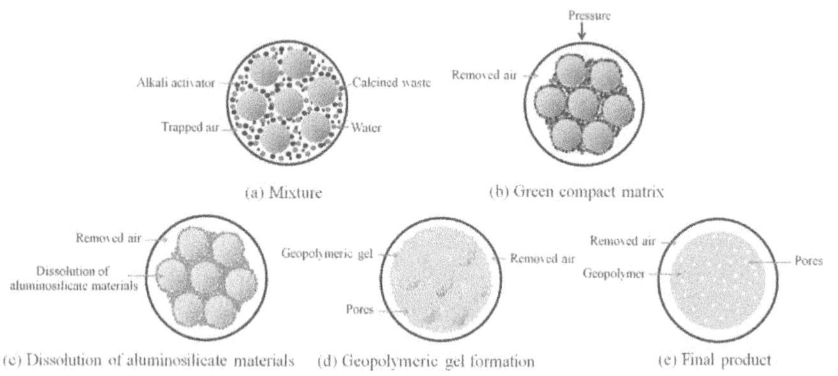

FIGURE 3.8 Schematic mechanism of geopolymer formation by the pressing method [2].

3.4.7 NMR Analysis

Green geopolymer prepared with calcined kaolin residue and low alkali activator content by pressing was synthesised with various Si/Al mole ratios using soluble silicate (Na_2SiO_3 solution) to improve the Si ratio. The structure of geopolymer with low alkali activator content was analysed by NMR, as shown in Figures 3.9 and 3.10 and Table 3.2. Findings found that when the geopolymer with Si/Al mole ratio was

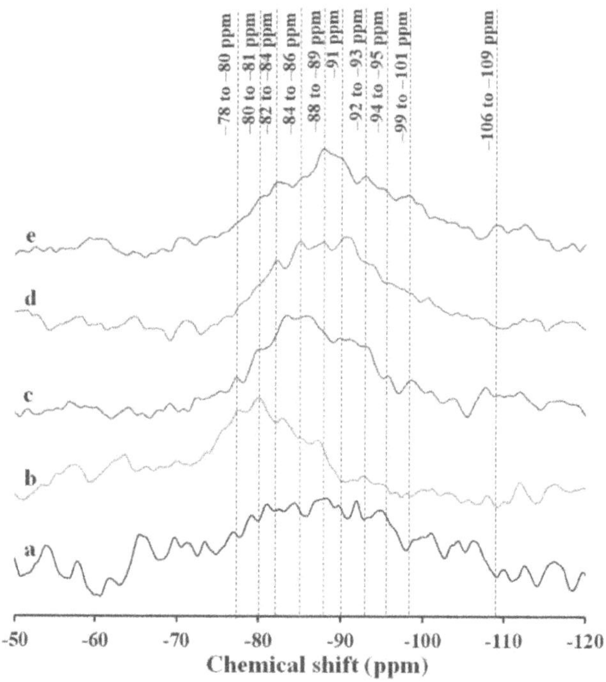

FIGURE 3.9 ^{29}Si NMR spectra of calcined kaolin residue (a) and geopolymer samples with Si/Al mole ratios of 1.50 (b), 1.54 (c), 1.58 (d), and 1.62 (e) [13].

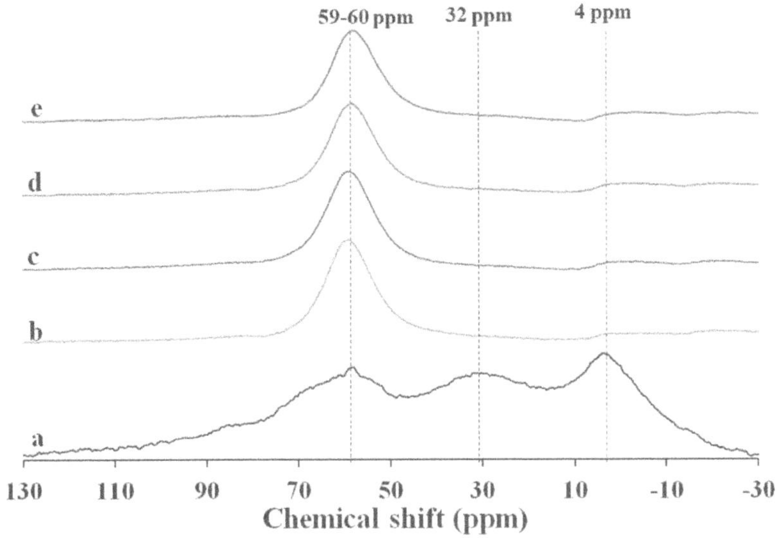

FIGURE 3.10 ^{27}Al NMR spectra of calcined kaolin residue (a) and geopolymer samples with Si/Al mole ratios of 1.50 (b), 1.54 (c), 1.58 (d), and 1.62 (e) [12].

increased to 1.58, Q4(4Al) and Q4(3Al) structures changed to Q4(2Al) structures, and Al^{IV} species declined. Consequently, the mechanical properties of the geopolymers may depend on the structure and chemical bonds within the three-dimensional aluminosilicate network. The higher compressive strength of the geopolymer was attributed to an increase in linkages of Si-O-Si bonds and larger aluminosilicate gels. However, the creation of the zeolite crystal phase (Figure 3.11) was responsible for an increase in the Si-O-Al and Al-O-Al bonds and a decrease in the gel aluminosilicate, which led to a reduction in compressive strength. The results show that the chemical

TABLE 3.2
^{29}Si and ^{27}Al NMR Spectra Deconvolutions (Area%) of Calcined Kaolin Residue and Geopolymer Synthesised with Various Si/Al Mole Ratios [13]

	^{29}Si NMR (%)									^{27}Al NMR (%)		
Si/Al Mole Ratios	Q^0	Q^1	Q^2 (1Al)	Q^3 (0Al)	Q^4 (0Al)	Q^4 (1Al)	Q^4 (2Al)	Q^4 (3Al)	Q^4 (4Al)	Al^{IV}	Al^{V}	Al^{VI}
1.42 (Calcined waste)	14.5	9.0	3.7	1.4	5.2	8.9	12.2	25.6	19.5	52.3	1	47.1
1.50	14.7	13.7	18.0	4.4	0.8	6.4	5.3	8.7	28.0	99.9	0	0.1
1.54	6.7	9.8	15.1	6.9	3.7	5.0	5.0	25.8	22.0	99.8	0	0.2
1.58	3.7	6.9	11.0	18.7	3.6	8.2	19.2	13.1	15.6	96.5	0	3.5
1.62	7.1	7.9	11.1	13.8	3.9	10.5	15.5	13.6	16.6	99.2	0	0.8

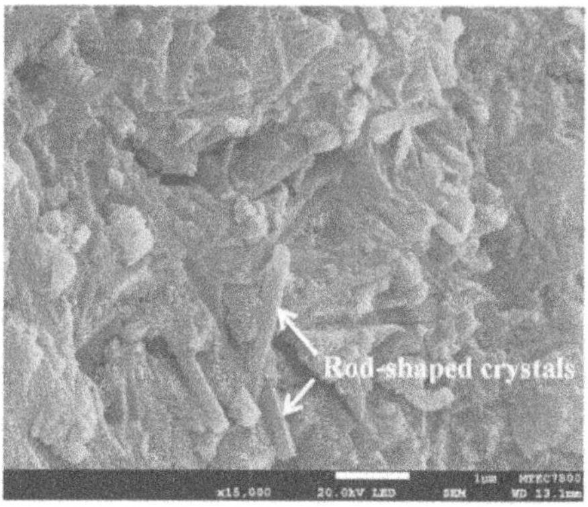

Elements	Wt%
O	48.46
Si	25.16
Al	15.74
Na	4.01
K	2.43
Fe	4.20
Total	100.00

a

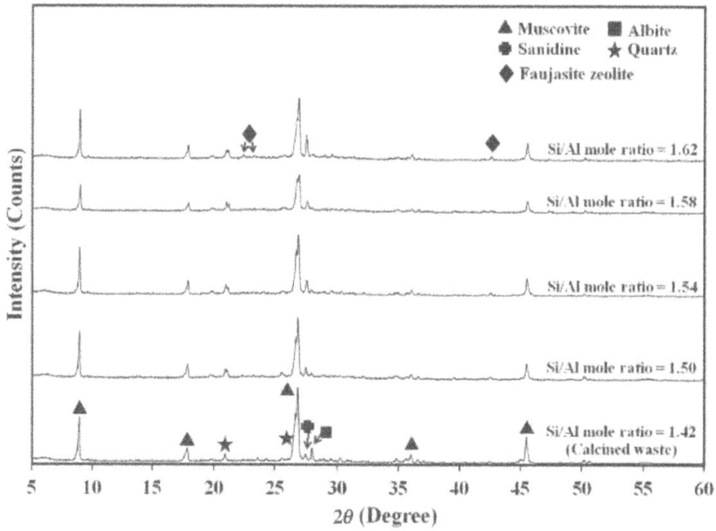

b

FIGURE 3.11 SEM image (a) and XRD diffractogram (b) of the zeolite phase in aluminosilicate gels of geopolymer with the highest Si/Al mole ratio of 1.62 [12].

composition of precursor materials affects geopolymers with low alkali activator levels in a substantial way. Additionally, the circular economy framework may be used to make geopolymers from kaolin residue and low alkali activator content, which will be used as green building materials to help with low carbon emission, low energy consumption, resource conservation, and environmental protection.

3.5 CONCLUSIONS

In this chapter, kaolin residue waste was utilised to synthesise geopolymers, with the waste being calcined before being used as a raw material. The following findings can be drawn:

1. The compressive strength and dense microstructure of geopolymer produced with a low amount of alkali activator solution by pressing was higher than that of geopolymer produced with a high amount of alkali activator solution by casting.
2. The geopolymer fabricated by the pressing method can reduce the amount of alkali activator solution by approximately 50% by weight compared to the casting method.
3. The optimum condition for the preparation of geopolymer was obtained from the pressing method, and a low alkali activator solution was found at 24 wt%.

REFERENCES

1. P. Kathirvel, S. Sreekumaran, Sustainable development of ultra high performance concrete using geopolymer technology, J. Build. Eng. 39 (2021), 102267.
2. S. Prasanphan, A. Wannagon, T. Kobayashi, S. Jiemsirilers, Reaction mechanisms of calcined kaolin residue-based geopolymers in the presence of low alkali activator solution, Constr. Build. Mater. 221 (2019) 409–420.
3. O. Shee-Ween, H. Cheng-Yong, L. Yun-Ming, M. M. A. Bakri Abdullah, H. Li-Ngee, P. Pakawanit, M. Suhaimi Khalid, W. H. B. Wan Muhammad, O. Wan-En, H. Yong-Jie, N. Yong-Sing, N. Hui-Teng, Green development of fly ash geopolymer via casting and pressing approaches: strength, morphology, efflorescence and ecological properties, Constr. Build. Mater. 398 (2023) 132446.
4. T. Alomayri, Experimental study of the microstructural and mechanical properties of geopolymer paste with nano material (Al_2O_3), J. Build. Eng. 25 (2019), 100788.
5. A.M. Rashad, M. Gharieb, Valorization of sugar beet waste as an additive for fly ash geopolymer cement cured at room temperature, J. Build. Eng. 44 (2021), 102989.
6. K.D. Wulandari, J.J. Ekaputri, S.B. Kurniawan, W.E. Primaningtyas, S.R.S. Abdullah, N.I. Ismail, et al., Effect of microbes addition on the properties and surface morphology of fly ash-based geopolymer paste, J. Build. Eng. 33 (2021), 101596.
7. S. Tuntachon, K. Kamwilaisak, T. Somdee, W. Mongkoltanaruk, V. Sata, K. Boonserm, et al., Resistance to algae and fungi formation of high calcium fly ash geopolymer paste containing TiO_2, J. Build. Eng. 25 (2019), 100817.
8. M. Alshaaer, Two-phase geopolymerization of kaolinite-based geopolymers, Appl. Clay Sci. 86 (2013) 162–168.
9. V. Živica, S. Balkovic, M. Drabik, Properties of metakaolin geopolymer hardened paste prepared by high-pressure compaction, Constr. Build. Mater. 25(5) (2011) 2206–2213.
10. S. Wang, X. Ma, L. He, Z. Zhang, L. Li, Y. Li, High strength inorganic-organic polymer composites (IOPC) manufactured by mold pressing of geopolymers, Constr. Build. Mater. 198 (2019) 501–511.
11. ASTM C109/C109-05 Standard Test Method for Compressive Strength of Hydraulic Cement Mortars (Using 2-in. or [50 mm] Cube Specimens) Annual Book of ASTM Standards, ASTM International, United States. (2008).
12. C. Tippayasam, P. Keawpapasson, P. Thavorniti, T. Panyathanmaporn, C. Leonelli, D. Chaysuwan, Effect of Thai Kaolin on properties of agricultural ash blended geopolymers, Constr. Build. Mater. 53 (2014) 455–459.
13. S. Prasanphan, K. Hemra, A. Wannagon, T. Kobayashi, S. Onutai, S. Jiemsirilers, ^{29}Si and ^{27}Al NMR study of the structural transformation of calcined kaolin residue-based geopolymer using low alkali activator content for sustainable construction materials, J. Build. Eng. 70 (2023) 106332.

14. K.L. Lin, H.S. Shiu, J.L. Shie, T.W. Cheng, C.L. Hwang, Effect of composition on characteristics of thin film transistor liquid crystal display (TFT-LCD) waste glass-metakaolin-based geopolymers, Constr. Build. Mater. 36 (2012) 501–507. doi:10.1016/j.conbuildmat.2012.05.018.
15. H.Y. Leong, D.E.L. Ong, J.G. Sanjayan, A. Nazari, The effect of different Na_2O and K_2O ratios of alkali activator on compressive strength of fly ash based-geopolymer, Constr. Build. Mater. 106 (2016) 500–511. doi:10.1016/j.conbuildmat.2015.12.141.
16. P. Duxson, J.L. Provis, G.C. Lukey, S.W. Mallicoat, W.M. Kriven, J.S.J. Van Deventer, Understanding the relationship between geopolymer composition, microstructure and mechanical properties, Colloids Surf. A Physicochem. Eng. Asp. 269 (2005) 47–58. doi:10.1016/j.colsurfa.2005.06.060.
17. B.H. Mo, H. Zhu, X.M. Cui, Y. He, S.Y. Gong, Effect of curing temperature on geopolymerization of metakaolin-based geopolymers, Appl. Clay Sci. 99 (2014) 144–148. doi:10.1016/j.clay.2014.06.024.
18. Z. Zhang, J.L. Provis, X. Ma, A. Reid, H. Wang, Efflorescence and subflorescence induced microstructural and mechanical evolution in fly ash-based geopolymers, Cem. Concr. Compos. 92 (2018) 165–177. doi:10.1016/j.cemconcomp.2018.06.010.
19. E. Najafi Kani, A. Allahverdi, J.L. Provis, Efflorescence control in geopolymer binders based on natural pozzolan, Cem. Concr. Compos. 34 (2012) 25–33. doi:10.1016/j.cemconcomp.2011.07.007.
20. S. Prasanphan, A. Wannagon, T. Kobayashi, S. Jiemsirilers, Microstructure evolution and mechanical properties of calcined kaolin residue-based geopolymers in the presence of different alkali activator content by pressing and casting, J. Met. Mater. Miner. 30 (2020) 121–132.

4 Effects of the Rheological Property on Extrusion Printing of Metakaolin-Based Geopolymers

Yanhong Jia and Huirong Le

4.1 INTRODUCTION

Geopolymers are low-carbon cementitious materials that hold promise to replace traditional concrete and enhance sustainability in the construction industry, especially when combined with 3D printing technology [1]. The geopolymer system varies according to the type of activator used, which can be an alkali, acid, or salt [2]. In particular, alkali-activated systems cure relatively quickly at room temperature [3]. Research on 3D printing of geopolymers has focused on linking material dynamic properties with the extrusion process. The proportions of geopolymer powder, activator, and rheological modifier can be adjusted to ensure 3D printing quality, at least in the short term [4]. However, changes in the rheological property during long-term continuous printing and high-precision printing may directly affect printing quality. If the rheological property during the printing process does not match the selected printing parameters, problems such as material interruption, nozzle clogging, and part collapse may occur [5]. Although rheological modifiers and fine aggregates can be added to reduce part deformation and collapse, the printing stability is still affected by the gel transition process, during which the viscosity, yield stress, and shear rate of the material undergo significant changes [6]. Thus, there is a high demand for an approach to optimizing 3D printing parameters to accommodate changes in the rheological property [7].

Current research on large-scale and long-term 3D printing of geopolymers for the construction field has been focused on alkali-activated geopolymers based on fly ash [8, 9]. Less research has focused on the 3D printing of alkali-activated geopolymers based on metakaolin, and any such studies have mostly considered smaller spatial scales [10]. Although substantial progress has been made in the material and process optimization of 3D printed geopolymers, the dynamic adjustment of printing parameters according to the rheological behaviour remains a critical issue to be addressed.

In this study, we developed a method for dynamically adjusting the printing parameters of alkali-activated metakaolin-based geopolymers in response to changes in the rheological property. We developed a Grasshopper-based G-code generation plugin that links key printing parameters such as the extrusion pressure, movement speed, and

DOI: 10.1201/9781003535683-4

extrusion height to real-time rheological property in the slurry, expanding the printable window and improving printing quality while reducing the need for additional additives. We developed a rheological model to determine the printable window for metakaolin-based geopolymer slurries with varying NaOH content, and we developed an extrusion model to describe the extrusion behaviour of the slurry. We conducted a series of experiments to verify the effectiveness of the proposed method and evaluate its impact on the printing quality. This study not only provides new insights for the development of 3D printing technology for geopolymers but also serves as a reference for the 3D printing of other materials that demonstrate changes in rheological property over time.

4.2 EXPERIMENTS AND METHODS

4.2.1 Material Preparation

We used an alkali-activated metakaolin material system to prepare the geopolymer slurry. The raw materials included metakaolin (98 wt%, Shanlin Shiyu Mineral Products Co., Ltd., Guzhang Xian, Hunan, China.), silica sol (Keying New Material Co., Ltd., Dezhou, China.), NaOH (96 wt%, Shanghai Titan Technology Co., Ltd., Shanghai, China), quartz sand (70–100 mesh, Qingmiaoyuan Co., Ltd., Zhengzhou, China), kaolin (95 wt%, Shanlin Shiyu Mineral Products Co., Ltd., Guzhang Xian, Hunan, China.), hydroxypropyl methylcellulose (HPMC-20, Qimeng Chemical Co., Henan, China.), and Triton X-100 (Scientific Phygene Co., Fuzhou, China.). Table 4.1 lists the chemical composition of the metakaolin used in this study, and Table 4.2 lists the physical parameters of the silica sol. Figure 4.1 shows the particle size distributions of metakaolin, kaolin, cellulose, and quartz sand.

The alkaline activator was prepared by mixing NaOH particles with 15 g of silica sol using a magnetic stirrer for 24 h. Then, the activator was mixed with 10 g of metakaolin powder and stirred for 10 min to obtain the cementitious material. Next, 1.5 g of quartz sand and 0.5 g of Triton X-100 were added along with appropriate amounts of kaolin and

TABLE 4.1
Chemical Composition of Metakaolin (wt%)

IL	Al_2O_3	SiO_2	Fe_2O_3	CaO	MgO	K_2O	TiO_2
0.41	45.03	52.14	0.48	0.13	0.15	0.14	1.16

TABLE 4.2
Physical Parameters of Silica Sol

SiO_2 Content (%)	Na_2O Content (%)	Density (g/cm³)	pH Level	Particle Size (nm)	Viscosity (mPa·s)
30 ± 1	≤0.3	1.19–1.21	9.0–10.0	8–15	≤10

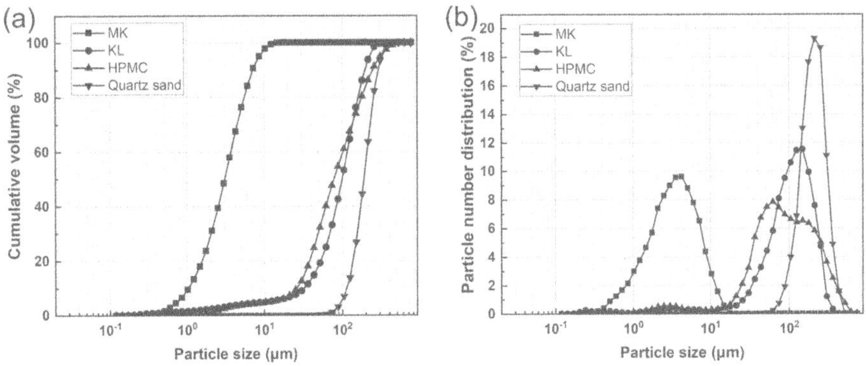

FIGURE 4.1 Particle size distributions of raw materials used in the geopolymer slurry: (a) cumulative volume and (b) particle number distribution.

HPMC as rheological modifiers. The mixture was stirred for another 10 min to prepare the geopolymer slurry, which was then loaded into the syringe of a modified direct ink writing extrusion head with a volume of 50 ml. Samples were labelled according to the amounts of NaOH, kaolin, and HPMC added. For example, sample Na5.33KL1.5HP1.5 contained 5.33 g of NaOH, 1.5 g of kaolin, and 1.5 g of HPMC.

4.2.2 3D Printer and Print Geometries

In this study, we modified a commercial ceramics 3D printer that can continuously print clay by pushing it through a lead screw extrusion head using steady pressure provided by an air pump. To address nozzle clogging caused by the curing of the geopolymer, we replaced the original lead screw extrusion head with a direct ink writing extrusion head, allowing it to be easily detached and replaced. The syringe of the extrusion head had an external diameter of 25.2 mm, an internal diameter of 22.5 mm, and a total length of 118 mm. The slurry was extruded through a needle with an internal diameter of 1.05 mm and a length of 20 mm. We added a pneumatic controller (DELIMA-983A) between the air pump and extrusion head to enable more precise dynamic pressure adjustment. Figure 4.2 shows the modified 3D printer and the associated equipment.

4.2.3 Material Design for Printability

To ensure that the cementitious material could be extruded through the 3D printer stably and smoothly, we needed to select suitable additives to modify the rheology of the cementitious material by enhancing its extrudability and buildability. Quartz sand was selected as the fine aggregate for improved buildability and reduced slump [11]. The ratio of coarse and fine sands was fixed for enhanced extrudability [12]. The layered kaolin structure had hydrophilic functional groups that reduced the free water content in the slurry and thus increased its viscosity and plasticity [13]. HPMC was selected because it can absorb and retain moisture, which would prevent cracking and shrinkage due to rapid drying [14]. Triton X-100 was selected as a surfactant to increase the yield stress of the slurry and facilitate smooth extrusion and rapid curing [15].

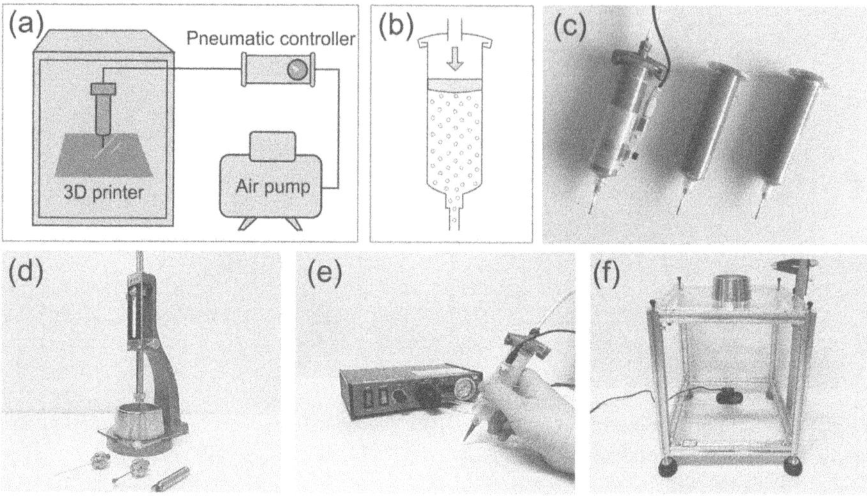

FIGURE 4.2 Modified 3D printer and equipment for testing the slurry printability: (a) Schematic diagram of key components of the 3D printer. (b) Schematic diagram of the direct ink writing extrusion head. (c) Replaceable direct ink writing extrusion head. (d) Vicat test equipment. (e) Pneumatic controller and direct ink writing extrusion head. (f) Mini slump test set-up.

4.2.3.1 Extrudability: Extrusion Test

A pneumatic controller and direct ink writing extrusion head (Figure 4.2c) were used to test the extrudability of the slurry samples with different proportions of NaOH, kaolin, and HPMC. The extrusion test started once the slurry was uniformly mixed and loaded into the extrusion head. The extrusion operation was performed every 15 min, and the extrusion line and extrusion pressure were recorded for each interval. Recording stopped when the maximum pressure of the air pump could no longer extrude the slurry, indicating a significant change in its rheological properties due to curing.

Slurry samples from the same batch were tested every 30 min at different extrusion pressures of 50, 100, 150, and 200 kPa. The extrusion volume was then calculated based on the dimensions of the extrusion head and the density of the prepared geopolymer (i.e., 2 g/cm^3). Each extrusion was performed for 10 s, and the weight of the extruded slurry was recorded by using an analytical balance.

4.2.3.2 Buildability: Mini Slump Test

Owing to the correlation between slump and rheological performance [16], we employed a self-built mini slump test (Figure 4.2f) to evaluate the relationship between the spread diameter of a printed slurry and time. The experiment was conducted on a transparent acrylic upper plate, with a camera installed beneath it to record the changes in spread diameter over time.

4.2.4 Rheological Performance

A dynamic shear rheometer (Model HAAKE MARS 60, Thermo Scientific, Beijing, China) was used to study the rheological performance of the printed slurries. The

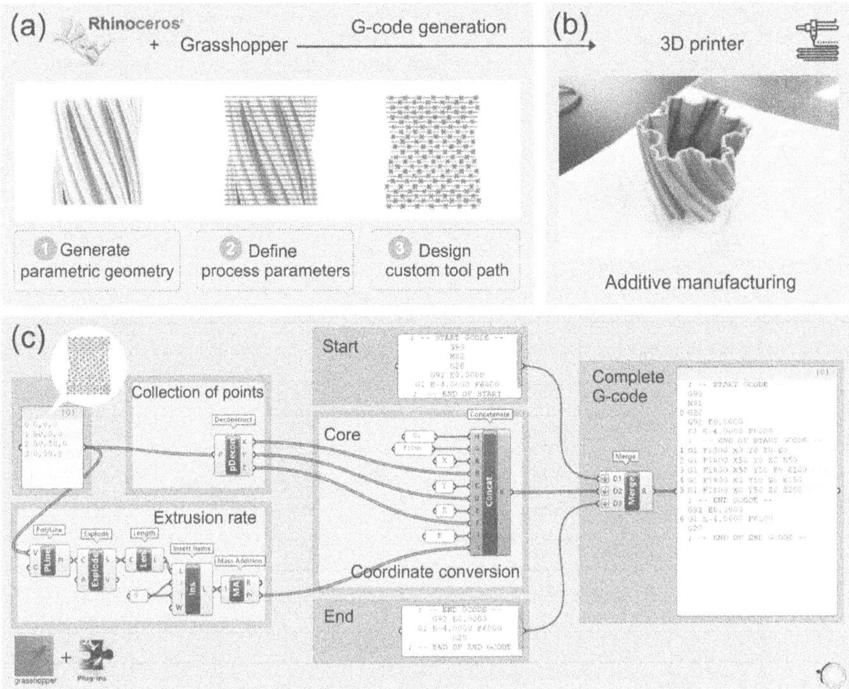

FIGURE 4.3 Grasshopper plugin for dynamic adjustment of printing parameters: (a) Example workflow of using the Grasshopper plugin to generate G-code files and print geopolymers directly. (b) Printing using the generated G-code files. (c) Screenshot of the software code implementing G-code conversion functionality.

shear rate sweep mode was applied to measure the variation of shear stress with shear rate, while the oscillation frequency sweep mode was used to obtain the changes in the storage modulus and loss modulus.

4.2.5 Software Development

To achieve dynamic adjustment of the printing parameters and more precise process control, we developed a plugin that closely integrates design, slicing, and manufacturing. We utilized the Grasshopper (Graphical algorithm editor) plugin to directly generate G-code files in the modelling software, Rhinoceros, and map dynamic changes in parameters to each coordinate point. Figure 4.3 shows the steps for generating the printing parameters and the interface of the developed plugin.

4.3 RESULTS AND DISCUSSION

4.3.1 Rheological Model Fitting

Experimental data of geopolymer slurries with varying NaOH concentrations were collected. As shown in Figure 4.4, data were obtained in shear scan mode (i.e., shear rate vs. shear stress) and oscillation scan mode (i.e., frequency vs. storage modulus

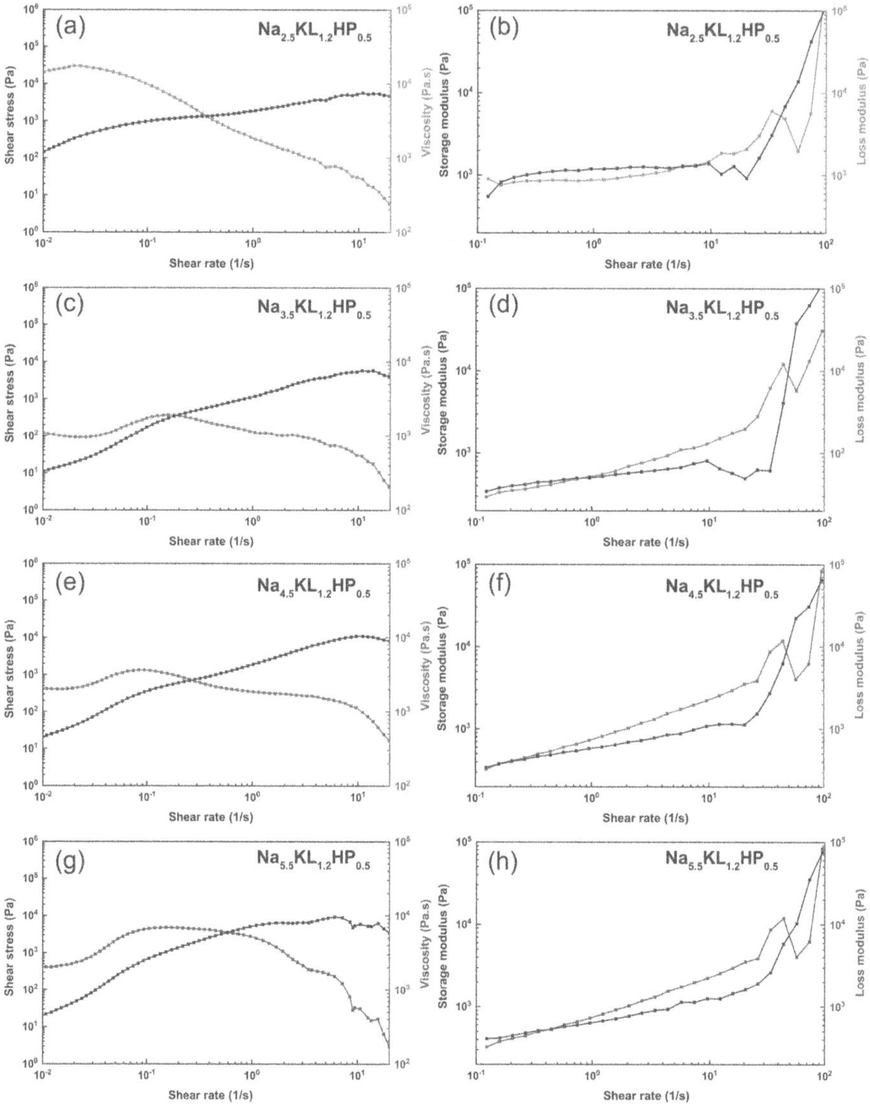

FIGURE 4.4 Experimental data for fitting to rheological models: (a, c, e, g) Variations in the shear stress with the shear rate of different samples. (b, d, f, h) Variations in the storage modulus and loss modulus with the oscillation frequency of different samples.

and loss modulus). To ensure reliability, the raw data scanned in the shear mode were preprocessed by setting the shear rate range to 0.01–20/s and filtering out extreme values.

We fitted the shear rate–shear stress data to three rheological models: the Bingham, Herschel–Bulkley, and Carreau models. The formulas, fitting parameters, and mean square error (MSE) for each model are detailed in the Supporting Information. Figure 4.5 shows the fitting curves of the three models.

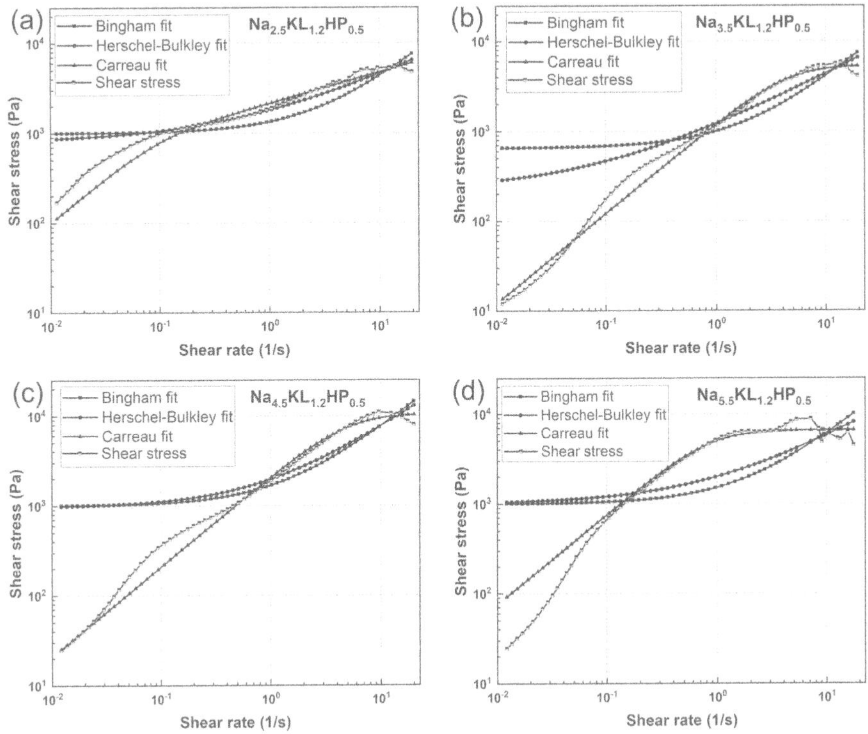

FIGURE 4.5 Shear rate–shear stress data of geopolymer slurries with different NaOH concentrations fitted to three rheological models.

For most samples, the Carreau model demonstrated the best fit and lowest error. The experimental results indicated that the prepared geopolymer slurries exhibit shear-thinning behaviour, as they showed a high initial viscosity at low shear rates, which gradually decreased as the shear rate increased [17, 18]. Therefore, the Carreau model was selected to investigate the laminar flow within the circular tube of the direct ink writing extrusion head.

4.3.2 Initial Setting Time

We measured the initial setting times of samples with different NaOH additions using the Vicat test apparatus (Figure 4.2d). The samples exhibited fluidity and shear strength in the initial setting state. As the fluidity decreased and viscosity increased, the penetration depth of the Vicat test needle gradually decreased until it could no longer penetrate the sample, indicating the final setting state. Figure 4.6a shows the initial and final setting times of the samples. Throughout the setting process, the yield stress changed significantly. The data obtained in oscillation scan mode (Figure 4.4b, d, f, and h) showed that both the storage modulus and loss modulus increased with an increasing oscillation frequency. However, the storage modulus increased at a greater rate than the loss modulus, which indicates that the material exhibited more elastic behaviour under periodic stress. This change can be attributed to the increased crosslinking density

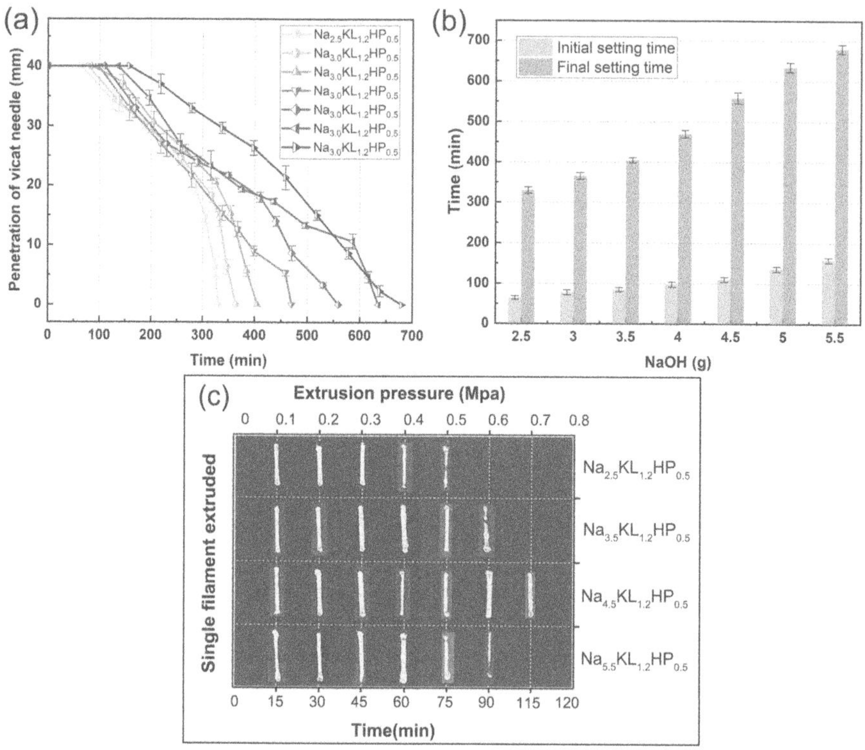

FIGURE 4.6 (a) Changes in penetration depth of the Vicat test needle over time. (b) Effects of the NaOH content in slurries on the setting time. (c) Extrudability test results of different slurries.

with a higher NaOH content. The addition of NaOH reduced the yield stress of the slurry, acting similarly to a superplasticizer [19]. The yield stress also varied owing to changes in the Na/Si ratio [20]. Increasing the NaOH content in the alkali activator delayed the initial setting time, which is mainly because NaOH reduces the polymerization of SiO_2 and slows down the setting speed [21]. An appropriate amount of NaOH helps increase slurry fluidity for smoother extrusion, but excessive NaOH can degrade the mechanical properties after curing [22]. The extrudability window for a slurry typically ends when the initial setting state is reached (Figure 4.6c). To minimize the use of NaOH and additives, optimizing the printing parameters is essential for extending the extrudability window of the slurry.

4.3.3 Direct Ink Writing Extrusion Model

The Benbow–Bridgwater model can be used to describe the stress and flow characteristics of geopolymer slurries extruded from a syringe, allowing for the calculation of extrusion air pressure and flow rate. The model is expressed as:

$$P_{ext} = \sigma_0 ln\left(\frac{A_b}{A_d}\right) + \frac{ML}{A_d}\left(\tau_0 + \beta V_{ext}^n\right) \tag{4.1}$$

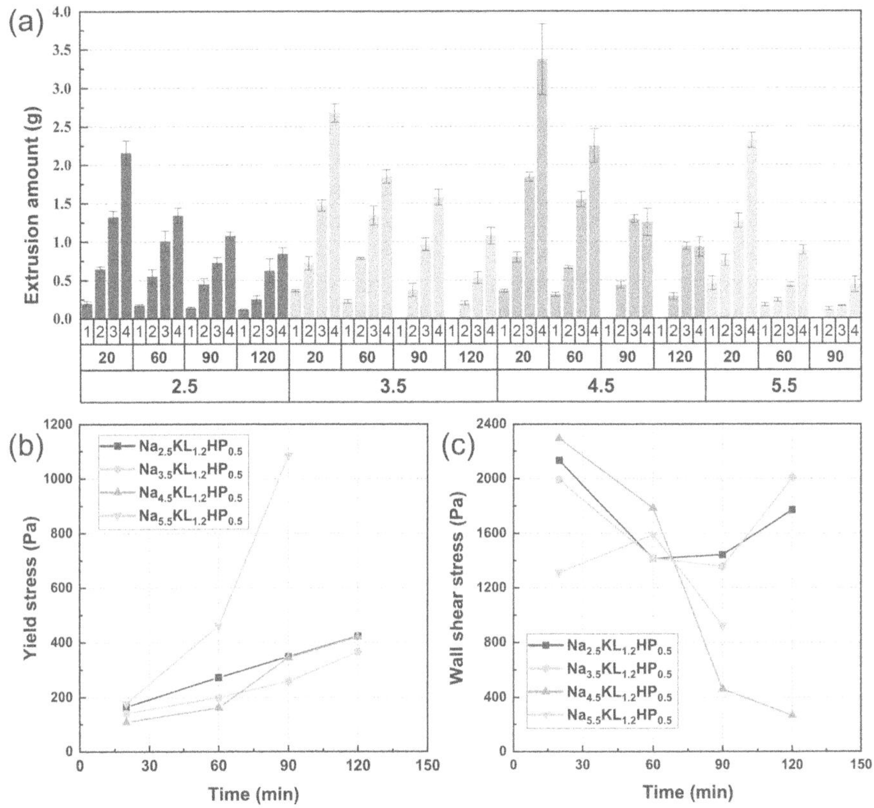

FIGURE 4.7 (a) Relationship between pressure and extrusion volume at different time points (20, 60, 90, and 120 min). On the x-axis, labels 1, 2, 3, and 4 correspond to extrusion pressures of 50, 100, 150, and 200 kPa, respectively. (b) Changes in yield stress over time. (c) Changes in wall slip shear stress over time.

Figure 4.7a presents data on the extrusion volumes of four samples at different time points (20, 60, 90, and 120 min) and various pressures. We used the extrusion data and Benbow–Bridgwater model to calculate the yield stress and wall slip shear stress of the four samples at different time points. The calculation process is detailed in the Supporting Information.

As shown in Figure 4.7b, the yield stress of each sample increased over time, with a significant influence from the NaOH concentration. As the NaOH concentration was increased, the yield stress initially decreased and then increased. At lower NaOH concentrations, more Si and Al could dissolve to produce more dissolved species [23]. As the NaOH concentration increased, the reactants quickly converted into N–A–S–H gel products, which gradually formed a denser microstructure, which increased the slurry viscosity and yield stress [24, 25]. As shown in Figure 4.7c, the wall slip shear stress initially decreased and then increased over time, which may be related to the shear-thinning characteristics of the slurries. In shear-thinning fluids, the viscosity decreases with an increasing shear rate, which affects the magnitude of the wall slip shear stress.

4.3.4 Dynamic Printing Parameter Settings

For successful 3D printing of geopolymers, the ideal rheological behaviour is a combination of a low yield stress and sufficient viscosity. This ensures that the slurry can easily extrude from the nozzle and stabilize quickly after deposition, avoiding deformation and collapse. Figure 4.8a shows the velocity and shear stress distributions within the syringe. The velocity distribution was parabolic along the diameter, while the shear stress was minimal at the centre and maximal at the inner wall. When the shear-thinning material reached a steady flow, the velocity along the diameter tended to be uniform, which is crucial for achieving even extrusion. At this stage, the viscosity of the material decreased with an increasing shear rate. The relationships between the shear stress, shear rate, and viscosity can be used to control the flow rate and avoid uneven extrusion or poor interlayer bonding.

Figure 4.8b shows the printing results of the slurry samples under different extrusion conditions: failed, unstable, and stable. In the failed state, the slurry collapsed or deformed. In the unstable state, uneven layering occurred. In the stable state, the slurry had uniform layer height and a smooth surface. During the 3D printing process, the printing parameters need to be adjusted dynamically based on the rheological properties and flow characteristics of the slurry within the syringe to account for changes over time, such as variations in layer height due to slump or discontinuous extrusion caused by insufficient pressure.

Figure 4.9 presents the 3D printing process and results after dynamic parameter adjustment based on the rheological properties and flow characteristics of the slurry within the syringe. The fitted data were input into the Grasshopper plugin to generate a G-code mapping the relationship between the extrusion pressure and the rheological performance. Figure 4.9a–c shows the printing results without dynamic parameter adjustment. Figure 4.9a shows the slump test results for slurry samples from the same batch, and Figure 4.9b shows the filling process of the direct ink writing extrusion

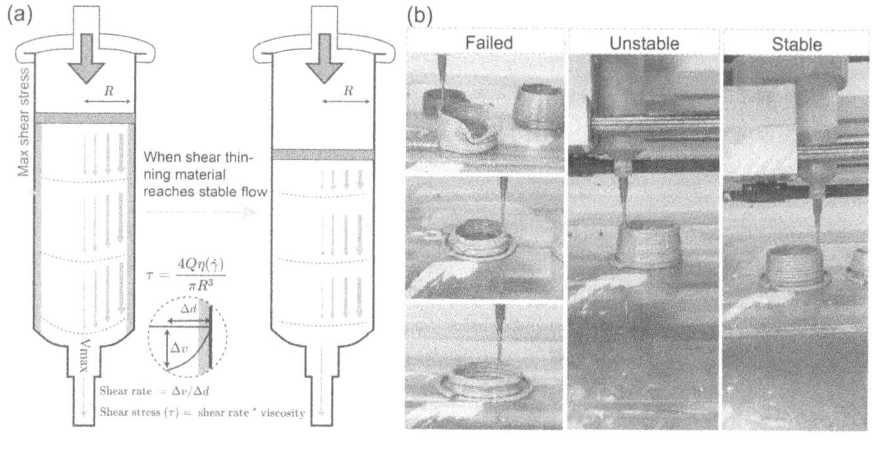

FIGURE 4.8 Flow velocity within the syringe and extrusion test results: (a) Schematic of the shear stress and velocity distributions. (b) Printing results under different extrusion conditions.

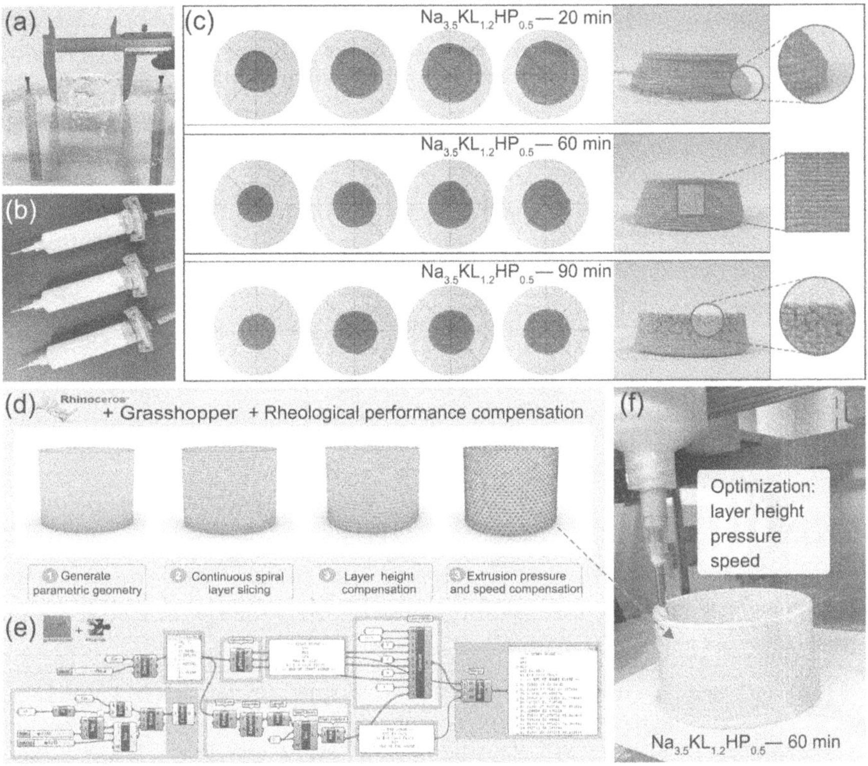

FIGURE 4.9 3D printing results of geopolymer slurries with dynamic adjustment of printing parameters.

head. Figure 4.9c displays the slump test and printing results after 20, 60, and 90 min of resting time, where material flowability decreased as the resting time increased, leading to issues such as uneven layer height, collapse, and increased surface roughness during printing. Figure 4.9d presents the rheological performance compensation method using the Grasshopper plugin in Rhinoceros. The plugin generates a parametric geometric model to achieve continuous spiral layer slicing, layer height compensation, and extrusion pressure and speed compensation for dynamic optimization of the printing process. Figure 4.9e shows the Grasshopper interface and the parameter adjustment process. Figure 4.9f illustrates the printing results obtained with dynamic parameter adjustment. Compared to the results without dynamic parameter adjustment, the obtained cylinder had a higher number of stacked layers, more uniform layer height, and reduced slump, indicating improved printing quality.

4.4 CONCLUSIONS

The 3D printing of alkali-activated metakaolin-based geopolymers requires careful consideration of rheological properties to ensure printing accuracy and stability. The experimental results indicated that dynamic adjustment of the printing parameters is crucial for improving the printing quality while minimizing the use of NaOH and rheological modifiers. The developed dynamic parameter adjustment method takes into account the changes in the material's rheological properties during the printing process to ensure the structural integrity and surface quality of the final printed product. This study provides important technical support for the application of alkali-activated metakaolin-based geopolymers in additive manufacturing and offers valuable insights and references for the 3D printing of other shear-thinning materials with real-time changes in rheological properties.

ACKNOWLEDGEMENTS

This work was supported by Tsinghua University Initiative Scientific Research Program-International Cooperation Special Project: 20233080011. The authors are also grateful for the support of colleagues at the Future Lab, Tsinghua University.

REFERENCES

1. S. Qaidi, A. Yahia, B. Tayeh, H. Unis, R. Faraj, A. Mohammed, 3D printed geopolymer composites: A review, Materials Today Sustainability 20 (2022): 100240, https://doi.org/10.1016/j.mtsust.2022.100240.
2. J. Davidovits, Geopolymers: Inorganic polymeric new materials, Journal of Thermal Analysis and Calorimetry 37.8 (1991): 1633–1656, https://doi.org/10.1007/bf01912193.
3. G. M. Zannerni, K. P. Fattah, A. K. Al-Tamimi, Ambient-cured geopolymer concrete with single alkali activator, Sustainable Materials and Technologies 23 (2020): e00131, https://doi.org/10.1016/j.susmat.2019.e00131.
4. H. Castillo, H. Collado, T. Droguett, S. Sánchez, M. Vesely, P. Garrido, S. Palma, Factors affecting the compressive strength of geopolymers: A review, Minerals 11.12 (2021): 1317, https://doi.org/10.3390/min11121317.
5. S. A. Khan, H. Ilcan, R. Imran, E. Aminipour, O. Şahin, A. A. Rashid, M. Şahmaran, M. Koç, The impact of nozzle diameter and printing speed on geopolymer-based 3D-printed concrete structures: Numerical modeling and experimental validation, Results in Engineering 21 (2024): 101864, https://doi.org/10.1016/j.rineng.2024.101864.
6. R. Aversa, L. Ricciotti, V. Perrotta, A. Apicella, Chemorheology of a Si/Al> 3 alkali activated metakaolin paste through parallel differential scanning calorimetry (DSC) and dynamic mechanical analysis (DMA), Polymers 15.19 (2023): 3922, https://doi.org/10.3390/polym15193922.
7. B. Panda, C. Unluer, M. J. Tan, Investigation of the rheology and strength of geopolymer mixtures for extrusion-based 3D printing, Cement and Concrete Composites 94 (2018): 307–314, https://doi.org/10.1016/j.cemconcomp.2018.10.002.
8. H. Alghamdi, S. A. Nair, N. Neithalath, Insights into material design, extrusion rheology, and properties of 3D-printable alkali-activated fly ash-based binders, Materials & Design 167 (2019): 107634, https://doi.org/10.1016/j.matdes.2019.107634.
9. Y. Chen, K. Xia, Z. Jia, Y. Gao, Z. Zhang, Y. Zhang, Extending applicability of 3D-printable geopolymer to large-scale printing scenario via combination of sodium carbonate and nano-silica, Cement and Concrete Composites 145 (2024): 105322, https://doi.org/10.1016/j.cemconcomp.2023.105322.
10. S. Ma, S. Fu, S. Zhao, P. He, G. Ma, M. Wang, D. Jia, Y. Zhou, Direct ink writing of geopolymer with high spatial resolution and tunable mechanical properties, Additive Manufacturing 46 (2021): 102202, https://doi.org/10.1016/j.addma.2021.102202.

11. J. Archez, N. Texier-Mandoki, X. Bourbon, J. Caron, S. Rossignol, Shaping of geopolymer composites by 3D printing, Journal of Building Engineering 34 (2021): 101894, https://doi.org/10.1016/j.jobe.2020.101894.
12. H. Zhong, M. Zhang, 3D printing geopolymers: A review, Cement and Concrete Composites 128 (2022): 104455, https://doi.org/10.1016/j.cemconcomp.2022.104455.
13. T. Voigt, J.-J. Mbele, K. Wang, S. P. Shah, Using fly ash, clay, and fibers for simultaneous improvement of concrete green strength and consolidatability for slip-form pavement, Journal of Materials in Civil Engineering 22.2 (2010): 196–206, https://doi.org/10.1061/(ASCE)0899-1561(2010)22:2(196).
14. B. Panda, S. C. Paul, M. J. Tan, Anisotropic mechanical performance of 3D printed fiber reinforced sustainable construction material, Materials Letters 209 (2017): 146–149, https://doi.org/10.1016/j.matlet.2017.07.123.
15. S. Ma, S. Fu, T. Yang, K. Li, G. Chen, Q. Dong, P. He, Z. Sun, X. Duan, D. Jia, P. Colombo, Y. Zhou, Unveiling the critical role of rheology modifiers in additive manufacturing of geopolymers and their mechanical properties, Additive Manufacturing 78 (2023): 103826, https://doi.org/10.1016/j.addma.2023.103826.
16. M. S. Choi, J. S. Lee, K. S. Ryu, K.-T. Koh, S. H. Kwon, Estimation of rheological properties of UHPC using mini slump test, Construction and Building Materials 106 (2016): 632–639, https://doi.org/10.1016/j.conbuildmat.2015.12.106.
17. N. S. Akbar, S. Nadeem, Carreau fluid model for blood flow through a tapered artery with a stenosis, Ain Shams Engineering Journal 5.4 (2014): 1307–1316, https://doi.org/10.1016/j.asej.2014.05.010.
18. L. Li, Y. J. Wei, Z. Li, M. U. Farooqi, Rheological and viscoelastic characterizations of fly ash/slag/silica fume-based geopolymer. Journal of Cleaner Production 354 (2022): 131629, https://doi.org/10.1016/j.jclepro.2022.131629.
19. K. Vance, A. Dakhane, G. Sant, N. Neithalath, Observations on the rheological response of alkali activated fly ash suspensions: the role of activator type and concentration. Rheologica Acta 53 (2014) 843–855, https://doi.org/10.1007/s00397-014-0793-z.
20. D.-W. Zhang, D. M. Wang, X.-Q. Lin, T. Zhang, The study of the structure rebuilding and yield stress of 3D printing geopolymer pastes, Construction and Building Materials 184 (2018): 575–580, https://doi.org/10.1016/j.conbuildmat.2018.06.233.
21. S. Chithiraputhiran, N. Neithalath, Isothermal reaction kinetics and temperature dependence of alkali activation of slag, fly ash and their blends, Construction and Building Materials 45 (2013) 233–242, https://doi.org/10.1016/j.conbuildmat.2013.03.061.
22. X. Luo, L. Huang, L. Wei, M. Chen, Z. Zhou, T. Zhang, A technique for preparing one-part geopolymers by activating alkali-fused lithium slag with solid sodium silicate, Construction and Building Materials 435 (2024): 136817, https://doi.org/10.1016/j.conbuildmat.2024.136817.
23. M. Nenadovic, M. Ivanovic, D. Kisic, N. Bundaleski, V. Pavlovic, S. Knezevic, L. Kljajevic, Changes in the physicochemical properties of geopolymer gels as a function of NaOH concentration, Science of Sintering 55.4 (2023): 509–519, https://doi.org/10.2298/sos220624020n.
24. B. Zhu, R. Yu, J. Shang, X. Xi, Y. Liu, X. Zhu, The influence of NaOH concentration on the strength and microstructure of ceramic binders prepared from coal gangue through geopolymerization, Materials Research Express 11.2 (2024): 025202, https://doi.org/10.1088/2053-1591/ad2a85.
25. K. Chen, W.-T. Lin, W. Liu, Effect of NaOH concentration on properties and microstructure of a novel reactive ultra-fine fly ash geopolymer. Advanced Powder Technology 32.8 (2021): 2929–2939, https://doi.org/10.1016/j.apt.2021.06.008.

5 3D Printing with Geopolymers and its Applications

Anton Frederik Becher, Szymon Gądek, and Kinga Korniejenko

5.1 INTRODUCTION

Additive manufacturing is a breakthrough technology in many areas, including the automotive industry, biomedical applications, and many others [1, 2]. Over the years, additive manufacturing (also often referred to as 3D printing) has grown from a niche method that was used almost exclusively for prototyping to a firmly established manufacturing technology. The process is characterised by its distinctive feature of building components from individual layers and, unlike conventional subtractive processes, using material only where it is needed in the product [3, 4]. The additive manufacturing technologies constitute a fast-developing industrial sector and, potentially, a groundbreaking technology. They have a lot of potential advantages for construction applications and can provide new horizons in this sector, i.e., increasing geometrical construction flexibility, reduction of labour costs, improvement of efficiency and safety, the possibility of construction in difficult environmental conditions, and they are in agreement with the sustainable development policy [4, 5]. It is worth noticing that this technology brings many benefits to building industries, one of them being automation and the possibility to work in harsh environments, including in space or underwater [1, 6]. However, the full exploitation of 3D printing technology for cementitious materials is currently limited [5].

Also, in recent years, there have been calls for more environmentally friendly production methods and materials for the building and construction industry. The most used building material worldwide is concrete, which has seen a sharp rise in demand over the last few years. The binder used for the concrete mixture is ordinary Portland cement (OPC), whose production is extremely energy-intensive and releases large quantities of carbon dioxide (CO_2) into the atmosphere [7]. To reduce emissions in this area, it is necessary to look for a more environmentally friendly binder that can be processed using modern production methods [8, 9].

Since their exploration in the 1970s, geopolymers, as a sustainable alternative, can reduce environmental impact due to their lower energy requirements and the ability to recycle industrial waste products [10, 11]. Geopolymers are mostly described as inorganic polymers formed by the reaction of aluminosilicate powders with alkaline solutions. Those preconditions result in materials that exhibit excellent mechanical

 DOI: 10.1201/9781003535683-5

properties, thermal stability, and chemical resistance [10, 12]. To fully utilise the potential of the material, modern production methods are required [13].

A promising solution in this scope seems to be geopolymers, but their application requires the development of the material and improvement of 3D printing technology for these materials. By leveraging the benefits of both geopolymers and 3D printing technology, this field holds promise for advancing sustainable manufacturing practices and developing next-generation materials for a wide range of industries [13, 14].

The motivation of this chapter is to show the importance of geopolymers as a printable material for civil engineering applications. The main aim of this study is to show the current development of geopolymer materials for 3D printing technology. The chapter is based on a literature review that is supplemented by case studies. The literature review is based mainly on the Scopus database. It was conducted between August and September 2024. The used keywords were 'geopolymer' and '3D printing' combined. The research results in the Scopus database showed 287 documents. The review was focused on the works for the last 5 years. The case studies include some projects conducted at the Faculty of Materials Engineering and Physics, Cracow University of Technology, Poland, or the project in which employees from this research unit were involved. The chapter is addressed to researchers and practitioners to demonstrate the state of the art in the field of application geopolymers as a material for 3D printing technology. It can be an inspiration for new works in this area.

5.2 3D PRINTING TECHNOLOGIES

The term 'Additive Manufacturing' is described in ISO/ASTM 52900 as a process of joining material. Parts from 3D model data are created by placing layer upon layer, as opposed to subtractive and forming manufacturing methodologies. Alternative names for additive manufacturing from a historical context are layered manufacturing and the term solid freeform fabrication. Most modern methods had their first appearance in the late 20th century, just like the techniques of laser photopolymerisation (Swanson, 1968), powder fusion (Ciraud, 1972), and laser sintering (Householder, 1979) [15]. The year 1987 can be seen as an important milestone in the history of AM with the appearance of the first commercially available stereolithography (SL) printer from 3D Systems [15]. In 2005, Dr. Adrien Bowyer presented his concept of RepRap (Replication Rapid-Prototyping), fuelling the success of a new, low-cost 3D printer, which provided many people with access to manufacturing [15]. Nowadays, we can distinguish between seven main methods of AM (ISO/ASTM 52900), which can be split into different sub-categories (Table 5.1) [16].

The ability to produce in such diverse ways enables the manufacturing method to establish itself in a variety of fields, including the automotive industry, building sector, medical sector, aerospace, defensive industries, and even the food market. Although additive manufacturing has a variety of methods and an almost unlimited choice of materials, the processes often follow the same procedure. For this reason, they can be generally divided into development, production, and post-production stages. Every process starts with a multidimensional model. This can be created in a CAD program through design, modelling, or by importing data from 3D scans, for

TABLE 5.1
Methods of Additive Manufacturing after ISO/ASTM 52900

Name	Description
Binder jetting	A liquid-based binding agent is used to selectively join material from a powder bed together [17, 18]
Material extrusion	Layered structures are created by the deposition of strands of material [19]
Powder bed fusion	It works by selectively adhering material in a powder bed, using a source of thermal energy [16]
VAT photopolymerisation	UV light is used to trigger polymerisation, which ultimately leads to the hardening of specific cross sections in a resin vat [17]
Direct energy deposition	Material in the form of powders or wire is fed into a laser or electron beam, which melts it and allows it to harden along the path [20]
Material jetting	Parts are created via selectively placing and curing photopolymeric droplets [17]
Sheet lamination	Parts are created from sheets of material which are stacked, then merged and cut to the desired shape [21]

example. In most cases, it is necessary to check the model before it can be optimised for additive manufacturing with the help of further programs.

The model can then be oriented and positioned in the virtual construction space in a slicer. The main task of the slicer is to convert the STL file into a G-code file. By horizontally dividing a closed three-dimensional body, a sequence of cross sections is created, whose geometries can be implemented as machine code. Furthermore, the exact printing parameters can be set, which plays an important role in the properties of the printed body. When the printer has material available and is at operating temperature, the printing process can begin. After the generated component has been 'unpacked', which often takes place in the form of separating it from the building platform or removing it from a bath or powder bath, the post-production stages follow. The post-processing can differ greatly in terms of the amount of work involved and account for a significant proportion of the overall process. Cleaning and support removal follow for most products. Frequently used processes are thermal and UV-based curing and classic mechanical and abrasive processes. There are also necessary post-processes, such as sintering or infiltration, as well as processes that can be easily automated, such as laser or plasma polishing, which are only suitable for certain materials [22, 23]. Visual and X-ray-based methods are usually used for quality control. In general, the workflow of additive manufacturing is shorter compared to conventional production methods. Especially since a lot of joining and processing steps have been eliminated [24, 25].

5.3 GEOPOLYMER AS A MATERIAL FOR 3D PRINTING

5.3.1 Key Material Properties

Geopolymers are convinced by a variety of qualifying properties, which are highlighted in the following section. Where it is suitable, its properties are compared with those of the world's most widely used construction material, that is, concrete [26, 27]. The setting of geopolymer (GP) happens between 35 and 600 min when being cured at 20–80°C. However, it can be accelerated or extended by customising the binder, composition, mixing process, or surrounding conditions [28]. When mixing geopolymers, water only fulfils the function of a chemical reaction medium, which means that it is not bound internally and causes the material to shrink up to 5% due to evaporation, which can lead to cracking. This can be prevented by using more fillers like sand or improving the molar ratios of reacting elements [29].

Most sources agree that shrinkage and weight loss happen more drastically in geopolymer concrete (GPC) compared to OPC concrete. Tukaziban *et al.* [30] compared a shrinkage of around 2.75% after the first 30 days in GPC, compared to only around 0.125% in OPC mixes [30]. The mechanical properties of geopolymer concrete strongly depend on the solution-to-binder ratio as well as the water content, the number of additives added [31], and the type of oxides used for providing SiO_2 and Al_2O_3 [32]. The values for flexural strength can reach up to 11 MPa, and a correlation between curing time and strength can be observed [33]. Bakri *et al.* [34] researched to compare the compressive strength of fly ash-based GPC against OPC concrete and found out that the latter achieved 37% lower values in testing [34]. It can be assumed that GPC offers the same or improved values as concrete with OPC [35]. According to data, geopolymers also have higher abrasion resistance (61% higher for a 12-h exposure and 64% higher for a 24-h exposure) [36] and rapid development of early strength (>35 MPa after 12 h) [37]. They also offer great bonding to reinforcing steel as they do not form different microstructures at the boundary layers [37].

The curing process, with regard to time and temperature, has great importance when it comes to explaining the properties of geopolymers. In most geopolymer mixes, curing at higher temperatures and for a longer time will lead to an increase in mechanical strength until reaching final properties [38]. Their resistance to acids and freeze-thaw circles due to the lack of water, as well as a low calcium content in some mixes, makes them a very durable material [38]. After being exposed to sulphuric acid for 18 months, blended-ash-based GPC specimens exhibited a mass loss of 8%, which is almost three times less than OPC, and only a 35% decrease in compressive strength values (compared to a 68% decrease in OPC) [39]. On top of that, they can absorb toxic chemical wastes and immobilise them by integrating and effectively sealing them inside the geopolymer matrix [40, 41].

The properties of geopolymers can be improved not only by varying the necessary components but also by adding additives. Nanomaterials such as silica fume, nano-ZnO, and graphene can improve the strength of geopolymers by up to 300% [42] (testing of flexural strength on geopolymer with graphene nanoparticles added) if used in very low percentages. The enforcement with fibres does significantly improve the compressive, tensile, and flexural strength and results in a less brittle material behaviour [43]. Their characteristics allow the use of numerous industrial waste materials as raw materials [44, 45] and make geopolymers very suitable for recycling [46]. Many articles attribute a significantly lower environmental impact to geopolymer binders compared to conventional/ordinary Portland cement. For example, Kumar and Kumar [47] describe geopolymers as materials that produce 4.8 times less CO_2 per ton in their production and only need 3/4 of the amount of water compared to OPC [47]. It must be pointed out that the sodium silicate solution has the worst environmental properties under the geopolymer ingredients [45].

5.3.2 Geopolymer Paste Properties

Additive manufacturing of a pasty material differs in many respects from working with conventional methods such as moulding. The properties of the paste must support uniform extrusion, good bonding to existing layers, minimal deformation after deposition, rapid curing, and other parameters. The flow behaviour of materials can be summarised under the term rheology. By testing the material with a rheometer, it can be divided into different fluid registers, each of which has different properties. Especially for systems in which the material is pumped, it is important that the paste behaves like an easily pumpable liquid under pressure and increases its viscosity as the stress decreases. This behaviour can be described as shear-thinning, and when the viscosity decreases not only under a shear strain but also as time progresses, we can speak about thixotropy. The activator solution plays a vital role in determining the viscosity of geopolymer mixes. Sodium-based activators lead to higher viscosities, and their molarity is an important regulator of flowability [48, 49]. Moreover, the activator should be coherent with the used raw material for geopolymer synthesis and its reactivity [50].

Another necessary property for clean processing is the good extrudability of the paste. A low plastic viscosity and optimal yield stress help to convey a material string that neither tears off nor leads to lumping. Zhang *et al.* [51] investigated the effect of the Si/Na ratio of the activator on the extrudability of geopolymer and found that decreasing this ratio leads to better results in terms of paste structure [51]. The workability of a material is characterised by how little resistance it offers to processing by conventional methods. This property is greatly influenced by the particle size and shape of the raw materials. Particle size reduction, for example, with fly ash generally improves the workability of the mixture, and the even particle shape also enhances machinability compared to slag-based geopolymer. The irregular particle shape and the compact structure of slag-based geopolymer lower its workability. It can be improved by replacing ground granulated blast furnace slag (GGBFS) with metakaolin to prevent early polymerisation and by adding additives just as superplasticisers to the mix [43, 52].

The open time of the material, i.e. how long it remains workable, and the buildability are closely related. Because fresh properties can change over time, it is important to know the time window during which the paste has optimal characteristics for extrusion. For geopolymers, the open time typically ranges up to 30 or 60 min [52]. However, variations in the precursors, temperature, and additives can have retarding or, conversely, accelerating effects on the material [7, 48].

Efforts in recent years to use materials more sparingly and to realise more sophisticated geometries, especially in the construction sector, require state-of-the-art production techniques. For geopolymers to establish themselves as a modern material, processability using additive manufacturing is of great importance. As with concrete, additive manufacturing of geopolymers ultimately promises the same improvements as conventional manufacturing methods [53, 54].

5.3.3 Extrusion Additive Manufacturing as the Most Popular Method for Geopolymer 3D Printing

With additive methods, geopolymers are mainly processed using extrusion-based techniques and less with the aid of binder jetting [55, 56]. Binder jetting utilises the powdered state of the geopolymer cement and solidifies them by adding the activator in droplets [57]. After analysing the essential parameters of the process, such as particle size, droplet distribution, and activator incorporation, Xia and Sanjayan [58] documented the mechanical and printing properties of binder jet-printed geopolymer geometries [58]. They used a powder mix made up of fine sand, anhydrous sodium metasilicate, and slag and cured the final printed part in sodium metasilicate solution to eliminate anisotropic properties [58, 59]. In another approach by researchers at ETH Zurich, a mix of metakaolin and silica sand was used as bed material for which a sodium silicate solution was printed. By changing the metakaolin content, they were able to establish correlations with powder wetting and binder distribution, as well as the optimum layer height [60].

Extrusion-based 3D printing methods require the presence of geopolymer in a mixed, paste-like form and convey it through a nozzle using force. A continuous material flow at the nozzle can be achieved by using a pump, a driven piston, or an extruder screw. The use of pumps or piston units for material feed is particularly suitable for larger printer systems. On the biggest scale, the nozzle is guided by a structural framework (gantry) or a robot arm. The gantry enables the construction of large cross sections in a cartesian coordinate system, whereas the robot arm works based on a polar coordinate system and is mostly used for slightly smaller prints. In both cases, the geopolymer mix is pumped through a hose before reaching the nozzle [61, 62].

In contrast, systems are also used where the material supply is positioned just above the nozzle, and a feed is achieved using extruder screws instead of direct force input. The principle is mainly used to develop and manufacture smaller components and has become established due to its ease of handling. The gantry systems have the advantage that they are basically an enlargement of most fused filament printers and are, therefore, working based on well-established principles [63] (Figure 5.1).

However, there are also differences between robots and gantry systems. Robotic systems have the great advantage that they are based on a 5-axis principle and can, therefore, also produce significantly more complex geometries, such as non-planar surfaces. Furthermore, they can be used more flexibly if not permanently mounted and enable higher print qualities. On the other hand, gantry systems are much easier to maintain and enable higher material throughputs.

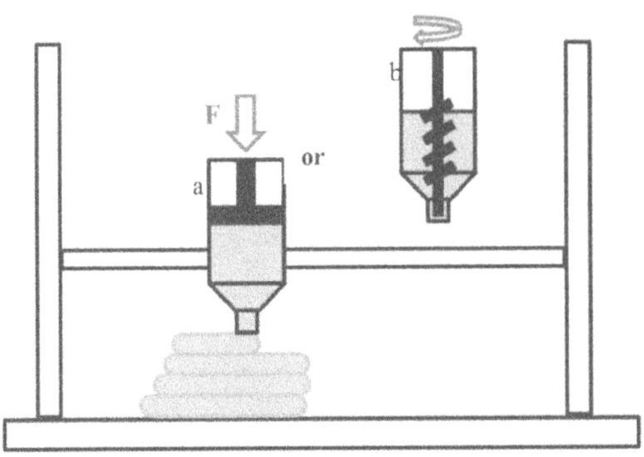

FIGURE 5.1 Illustration of a paste extrusion printer based on a 3-axis frame with material feed by (a) pressure and (b) screw.

Nowadays, practical tests in the area of geopolymer materials are usually carried out with the help of the paste extrusion modelling (PEM) method. It is one of the extrusion-based methods and is often used synonymously with liquid deposition modelling (LDM). LDM was shaped by the Italian company Wasp, which used the term for its special extrusion process for ceramic and clay-based materials. The principle is based on feeding a pasty material through a nozzle and depositing it at a defined position. The advantages of these processes are that very large components can be generated quickly and inexpensively from a wide range of materials. However, the geometric accuracy can vary greatly due to the nozzle diameters and layer thicknesses [64, 65].

5.3.4 The Main Area of Current Applications

The largest use of additively processed geopolymer is in the construction industry, where it can complement the possibility of modern construction using 3D printing by utilising a more environmentally friendly material in comparison to OPC [66, 67]. The initial state of research in the field of AM with geopolymers results in only a few examples of applications (Figure 5.2).

Nematollahi *et al.* [71] investigated the printability and testing of mechanical properties for a 'one-part' geopolymer, which can be used in 3DP applications after mixing with water. They utilised a gantry and auger-type extruder to generate defined specimens and large free-form shapes (Picture A) [71]. To enhance the low yield stress of geopolymers, Panda *et al.* [68] introduced nano clay to their geopolymer concrete and managed to manufacture a small-scale bathroom unit, proving the workability and extrudability of their composition (Picture B) [69].

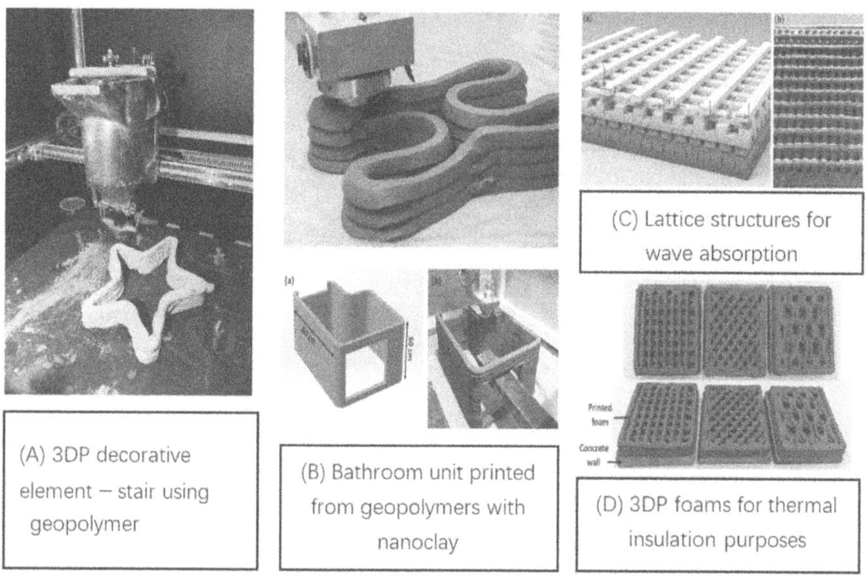

FIGURE 5.2 Different applications for AM with geopolymers [68–70].

Picture C shows wave-absorbing lattice structures printed using the dual gradient direct ink writing (DGDIW) method through the continuous transition from one material to another. The specimen displayed a largely increased absorption of electromagnetic waves, and the combination of materials and methods achieved highly functional results [69]. Alghamdi and Neithalath [70] used paste extrusion to generate geopolymer foams with enhanced thermal insulation properties. Their porous material is tested for its mechanical, conductive, and thermal properties and then evaluated on the basis of already existing materials (Picture D) [70].

It is also worth mentioning that 3D printing also uses geopolymer materials [72, 73]. They are usually made by utilising a large gantry system, and they were able to print a geopolymer mix based on liquid silicates combined with reactive materials as well as sand because the complicated mixing and preparation of the geopolymer material remains challenging [74, 75].

Another field of application for printed geopolymer geometries is artificial coral reefs and similar underwater applications. Practical testing in this area was done by A.I. Yoris-Nobile *et al.* [76]. In various research studies, the effects of material and design were tested in terms of environmental protection, cost, impact, and embedding in nature. After analysing different geometries of geopolymer and Portland cement in differing marine regions and for varying periods of time, similar properties were found; however, Portland cement provided better mechanical values [76, 77].

5.4 CASE STUDIES OF THE DEVELOPMENT OF 3D PRINTING TECHNOLOGY FOR GEOPOLYMERS

5.4.1 Urban Infra Revolution Project

The project 'Urban infra revolution: Circular economy materials and the development of novel methods to produce recyclable and functional urban construction products' was co-financed by the European Regional Development Fund through Urban Innovative Actions (UIA). The project was implemented between 2017 and 2020 by the consortium, which involves the municipal – city of Lappeenranta, Finland (lider), four small-medium enterprises (SME), five private enterprises (large), two universities, and one Region Development Company (Table 5.2).

This project involves several activities connected with sustainable development and circular economy. The most important was the valorisation of industrial by-products and waste for geopolymer manufacturing and the implementation of 3D printing technology to reduce CO_2 emissions, limit the amount of waste during production, and increase cost efficiency. The new geopolymer materials are based on by-products from local industry, such as ashes, green liquor dregs, and mine tailings, as well as on construction and demolition waste. These materials are produced in large amounts in Finland, and previous research shows the usefulness of such kinds of products in the geopolymerisation process [78, 79]. The main idea connected with the material design was to limit environmental burden through the replacement of concrete in urban architecture. Because of that, the designed geopolymer composite material includes 99.6% of circulative materials and is 100% recyclable. Additionally, the composite is based on local sources (a maximum distance of 100

TABLE 5.2
The Most Important Information about the Project Urban Infra Revolution

Project title	Urban Infra Revolution – Circular economy materials and novel method development to produce recyclable and functional urban construction products
Amount and sources of funding	EU: co-financed by the European Regional Development Fund through the Urban Innovative Actions (UIA) Total project: 4 336 568.40 EUR Co-financing: 3 386 638.40 EUR
Duration	01/11/2017–31/12/2020
Consortium	Lider: Municipal—City of Lappeenranta (Finland) Partners: 4 SME, 5 private enterprises, 2 universities, 1 Region Development Company
Raw materials	Ashes, green liquor dregs, tailings, construction waste
Technology	Laboratory (testing materials): Robotic arm; Scale-up: Large-format printer – printing inside the ready elements
Product	Noise barriers, small architectural elements such as pots and benches, skate park

km from the manufacturing plant), which realistically reduces the cost of transportation and CO_2 emission. Additionally, the materials had to be coherent with additive 3D printing technology.

The application of 3D printing technology for the geopolymers was an important innovative aspect. The challenges were connected with automation of the technology (efficiency, zero waste) and using it in harsh environments (low temperatures in Finland). In the laboratory, for the testing materials, the robotic arm was used (Figure 5.3a). For scale-up, a new solution – a large format 3D printer – was developed based on the solution for the concrete 3D printing (Figure 5.3b). The prototype elements were performed using a large 3D printer (Figure 5.3c). These products were dedicated to urban architecture, including elements for a skate park and a noise barrier. The design process was performed with applications of new technologies (3D modelling and augmented reality). The final elements were characterised by high aesthetic values thanks to the use of additive manufacturing solutions. One of these products, a noise barrier (100 m length), is presented in Figure 5.3d.

The project experience shows that the mine tailings or by-products from the industry are potentially useful for the geopolymerisation process and for the production of valuable products for urban architecture. The described case study indicates an alternative to the current methods of waste rock management. Also, it proves the usefulness of 3D printing technology for design elements for urban architecture, giving them interesting aesthetic values as well as some new properties, such as an increased level of sound absorption because of the rough surface.

5.4.2 PRINTGEOHOUSE Project

The project titled 'Development of 3D printing technology for construction and facade prefabricated elements made of concrete composites and geopolymers' was supported with European funds dedicated to the development of the Polish economy (Table 5.3).

FIGURE 5.3 Photos from the project realisation: (a) 3D printing on a small scale using a robotic arm, (b) 3D printer constructed for the large elements manufacturing, (c) ready elements, and (d) final product – noise barrier.

One of the requirements of such grants was cooperation with the company and receiving the final stage of the advanced prototype solution for practical application.

The main aim of the activities was to obtain the building elements for residential houses made using 3D printing technologies. Two different types of materials were applied to this purpose: concrete composites and geopolymers, including hybrid solutions. In the first stage, manual trials in the laboratory were conducted using

TABLE 5.3
The Most Important Information about the Project PRINTGEOHOUSE

Project title	Development of 3D printing technology for construction and facade prefabricated elements made of concrete composites and geopolymers (PRINTGEOHOUSE)
Amount and sources of funding	National (Polish)/EU – Polish National Centre for Research and Development in Poland Total project: ca. 1 784 050.00 EUR Co-financing: ca. 806 342.38 EUR Coordinators: Prof. Janusz Mikuła, Prof. Marek Hebda, CUT
Duration	01/01/2019–31/12/2021
Consortium	Cracow University of Technology, Poland (CUT) CKBM Sp. z o. o. Sp. K. (Company), Poland
Raw materials	Metakaolin, fly ash
Technology	Laboratory (testing materials): modified WASP 2040; Scale-up: Large-format printer (ATMAT)
Product	Components for the production of a residential house on the place

FIGURE 5.4 Photos from the project realisation: (a) used 3D printer, (b) problems with the liquidity of geopolymer paste, and (c) 3D printing process for exemplary element.

geopolymers and their composites reinforced by fibres [80, 81]. For the final stage, the decision about the application of short fibres was made because of the easier practical application of 3D printing technology [82]. In the next phase of research, the 3D printer was applied to manufacturing some elements (Figure 5.4a).

During the process, some challenges have been noted, including the proper proportion of the liquid-solid ratio (Figure 5.4b and c). Moreover, it was noticed that this proportion is dependent on the scale of 3D printed products, and material behavior is different on a small scale than on a larger scale. Another important conclusion was the strong dependence of the quality of final products on the stabilisation of the pumping process. The brakes in this process caused the failures in the final products. During the project, some ideas connected with heating elements during the 3D printing process were also tested. Two methods were developed: direct heating of paste after realising the nozzle and 3D printing on the heated place. Both ideas significantly complicated the process of manufacturing, but at the same time, they gave poor results and caused problems with process stabilisation. Finally, the process was made at ambient laboratory temperature without an additional heating system. In this case, the most promising solutions were obtained based on hybrid materials (concrete and geopolymer) reinforced by fibres [83, 84]. The provided works show a high dependence on mechanical strength depending on the direction of 3D printed layers. The obtained materials in full scale were weaker than elements made with casting technology; however, the properties that were achieved were sufficient for applications in residential buildings [84].

5.4.3 Microstructure Investigation

The project under the title: 'The influence of the material structure on the mechanical properties of geopolymer composites reinforced with short fibres obtained with additive technologies' has a supporting character to recognise the phenomena inside the material and their influence on material properties (Table 5.4).

The main goal of the project was to investigate the influence of material microstructure on the mechanical properties of the two kinds of composites (flax and

TABLE 5.4
The Most Important Information about the Project MINIATURA 3

Project title	The influence of the material structure on the mechanical properties of geopolymer composites reinforced with short fibres obtained with additive technologies
Amount and sources of funding	PL: National Science Center Poland, MINIATURA 3 Total project: 6 500 EUR Coordinator: PhD Eng. Kinga Korniejenko
Duration	22/11/2019--21/11/2020
Consortium	Cracow University of Technology, Poland (without partners)
Raw materials	Fly ash with fibre additions
Technology	Manual trials (extrusion)
Product	3D printable geopolymer reinforced by fibres

carbon reinforced) and compare two methods of production for geopolymer composites (casting and 3D printing). The results partly explain these phenomena and help better understand the relationship between microstructure and mechanical properties of the composites. The results also provided the possibility of comparing different manufacturing methods. The samples made by the additive technology had comparable properties with those made by casting. The main finding in this research was the very high values of compressive and flexural strength of flax fibres in comparison with carbon fibre-reinforced composites. It confirms the possibility of usage as a reinforcement in the geopolymer matrix with natural fibres [85].

5.4.4 Smart Geopolymers

The main area of the 'Smart Geopolymers' project was the valorisation of the mining products. Special attention was given to by-products from the mining industry, especially aluminium production (bauxite residues) [86]. The important goal of this project was to prepare a material that would be printable and fire-resistant. The main idea for the application of this material was the internal layer of the tunnels. Because of that, it was necessary to obtain a significant level of freedom of shaping and, for safety reasons, fireproof properties (Table 5.5).

The material composition was prepared mainly by partners from Belgium and Portugal, based on raw products delivered by Greek partners and using a geopolymerisation process. The main area of Polish partners was the stabilisation of the process of 3D printing. The obtained material has very good properties for that purpose, especially pumpability and viscosity. The material received good mechanical properties at ambient temperature after several days. It was possible to shape it into complex products (Figure 5.5).

The important aspects of the project were also environmental issues. The combination of sustainable material with 3D printing technology allows the design of the technology that brings very clear environmental benefits. It not only re-uses mining by-products but also, thanks to the usage of 3D printing technology, fulfils the requirement for zero waste production with a significant level of energy efficiency.

TABLE 5.5
The Most Important Information about the Project SMART-G

Project title	Smart Geopolymers (SMART-G)
Amount and sources of funding	EU: ERA-MIN 2 (Call 2019), TOPIC 4. Recycling and Re-use of End-of-Life products Total project: 1 085 926 EUR (Coordinator: Prof. Hubert Rahier, VUB) CUT: 116 250 EUR (Coordinator CUT: Prof. Izabela Hager)
Duration	01/12/2020–31/11/2023
Consortium	Lider: VUB – Vrije Universiteit Brussel, Belgium Partners: Portugal: University of Aveiro Greece: MNLT Innovations GP, IESL/FORTH, Mytilineos S.A. Poland: Cracow University of Technology, PBP Łęgprzem Sp. z o.o, Poland Belgium: ResourceFull
Raw materials	Industrial waste, red mud, CDW, etc.
Technology	Extrusion – ATMAT company
Product	Tunnel

5.4.5 Development of Lunar Regolith Simulant for 3D Printing in Binder Jetting Technology

The project titled 'Development of lunar regolith simulant for 3D printing in Binder Jetting technology' was a small grant, but it is worth mentioning because of the very interesting topic and experience connected with this activity can be used for designing the shelters on the Moon [87] (Table 5.6).

The main goal of the project is to develop materials based on ceramics for production using 3D printing technology – Binder Jetting. The developed material (powder) is intended to imitate the lunar regolith (soil) in terms of its chemical composition; an alkaline material, previously used in geopolymer materials, is planned as the binder material. The implementation of design tasks allowed for the expansion of knowledge

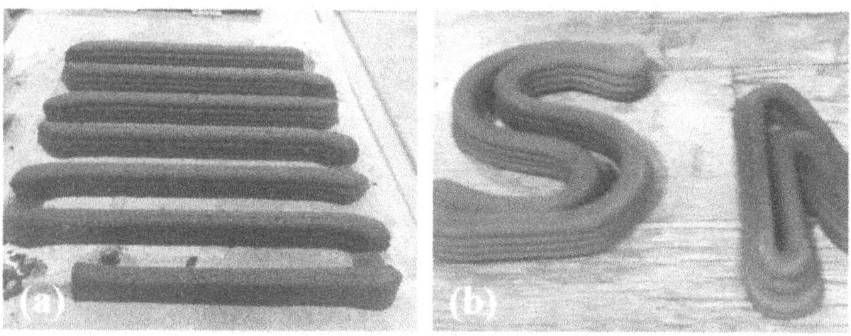

FIGURE 5.5 Photos from the project realisation: (a) 3D printed material for samples and (b) exemplary elements – letter.

TABLE 5.6
The Most Important Information about the Project MINIATURA 5

Project title	Development of lunar regolith simulant for 3D printing in Binder Jetting technology
Amount and sources of funding	PL: National Science Center Poland, MINIATURA 5 Total project: 6 000 EUR Coordinator: PhD Eng. Barbara Kozub
Duration	02/12/2021–01/12/2022
Consortium	Cracow University of Technology, Poland (without partners)
Raw materials	Selected powders for regolith composition
Technology	Manual trials (binder Jetting)
Product	3D printable regolith (simulant of lunar regolith)

regarding the possibilities of new applications of materials, both as in-situ raw materials for the construction of the lunar base, as well as for 3D printing in extreme conditions on the Earth, and the use of geopolymer materials for special applications in the construction industry. One important element of the works was the research on the microstructure of composites, including the morphology of both ceramic and metallic powder particles and their impact on the durability of joints and final material properties. The use of binder jetting technology was connected with the most probable solution possible to apply in the outer terrestrial environment [87].

5.4.6 3D-FOAM Project

The ongoing project connected with the 3D printing application is conducted under the title 'Foamed Geopolymer Made by Additive Manufacturing for the Construction Technology Applications (3D-FOAM)' (Table 5.7).

The main aim of this project is to print the foamed material that can be used for insulation purposes in buildings. The first trials show that this kind of task is quite ambitious and caused some challenges connected with proper foam formation and stabilisation. To obtain the proper solutions, proper additives must be used. The most valuable results have been achieved with silica fume and aluminium for foam formation and polymeric fibres for stabilisation. The slightly worse results have been acquired by using hydrogen peroxide [88]. To better control the 3D printing process, the trials are made on the 3D printer that has a chamber with temperature regulation (Figure 5.6).

An important challenge undertaken in the project is the design of zero-emission technology for 3D printing and the use of waste products as raw materials, such as clay bricks, aerated concrete, crushed cement, etc. However, the most important innovative area is in the development of foamed geopolymer composites for 3D printing with tailored properties.

5.4.7 MAR-WRECK Project

An ambitious task was undertaken in the project MAR-WRECK: 'Development of geopolymer composites as a material for protection of hazardous wrecks and other critical underwater structures against corrosion'. In this case, the trials of underwater

TABLE 5.7
The Most Important Information about the Project 3D-FOAM

Project title	Project: Foamed Geopolymer Made by Additive Manufacturing for the Construction Technology Applications (3D-FOAM)
Amount and sources of funding	EU: M-ERA.NET 3 Call 2021 Total project: 757 000 EUR Co-financing: 210 000 EUR Coordinator: PhD Dariusz Mierzwiński
Duration	01/05/2022–30/04/2025
Consortium	Lider: Cracow University of Technology, Poland Partners: ATMAT Sp. z o. o., Poland Riga Technical University, Latvia University of Miskolc, Hungary National Ilan University, Taiwan
Raw materials	Ashes, tailings, others Hybrids with concrete
Technology	Extrusion – dedicated equipment (ATMAT, PL)
Product	Insolation materials for the construction industry

3D printing were undertaken (Table 5.8). It was a very interesting task because exploring underwater areas is considered an even more challenging task than designing the materials for space applications. It is worth stressing that, in many ways, we know more about the surface of our moon than we do about the extreme depths of our oceans. More people have walked on the moon than have reached the bottom of the Mariana Trench, more on which later [89].

Basically, this project is a response to the need for materials with increased durability to protect dangerous wrecks and critical underwater infrastructure against

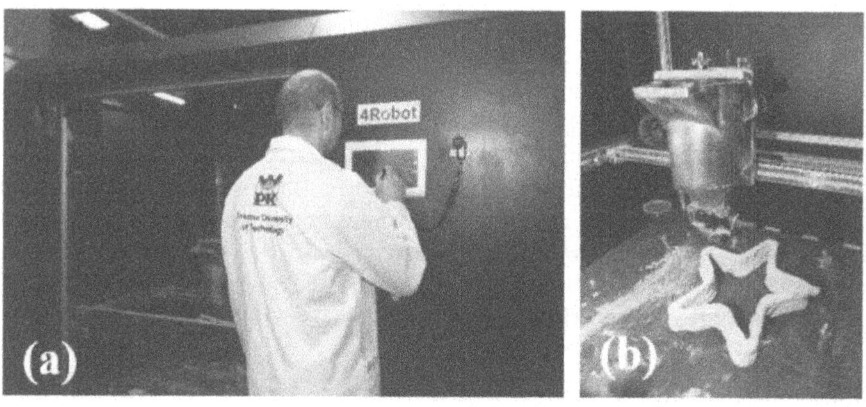

FIGURE 5.6 Photos from the project realisation: (a) programming of 3D printer and (b) nozzle and exemplary element.

TABLE 5.8
The Most Important Information about the Project MAR-WRECK

Project title	Development of geopolymer composites as a material for protection of hazardous wrecks and other critical underwater structures against corrosion (MAR-WRECK)
Amount and sources of funding	EU: M-ERA.NET 3 CALL 2021, Call Topic: High performance composites Total project: 1 141 860 EUR (Coordinator: PhD Thomas Grab) CUT: 275 000 EUR (Coordinator CUT: PhD Kinga Korniejenko)
Duration	01/06/2022–31/05/2025
Consortium	Lider: Technische Universität Bergakademie Freiberg, Germany Partners: Cracow University of Technology, Poland HIBRID Sp. z o.o, Poland Technical University of Liberec, Czech Republic
Raw materials	Ashes, tailings, construction waste, etc.
Technology	Extrusion and binder jetting
Product	Material for underwater applications

corrosion. As part of the project, it is planned to find a cost-effective method that will protect shipwrecks, particularly wrecks containing cargo that may be potentially hazardous to the environment. The development of such a method will eliminate the risk of leakage and release of hazardous substances, such as petroleum waste, into the natural environment. The material that has been developed for this purpose is geopolymer composites because of their resistance to marine environments [90, 91].

For this ambitious task, 3D printing technology seems to be an adequate solution. It allows the design of the proper elements for the protection of wrecks according to the specific shape of each of them. Part of this project was to do trials of underwater 3D printing and the development of the material for 3D printing in the water environment (Figure 5.7).

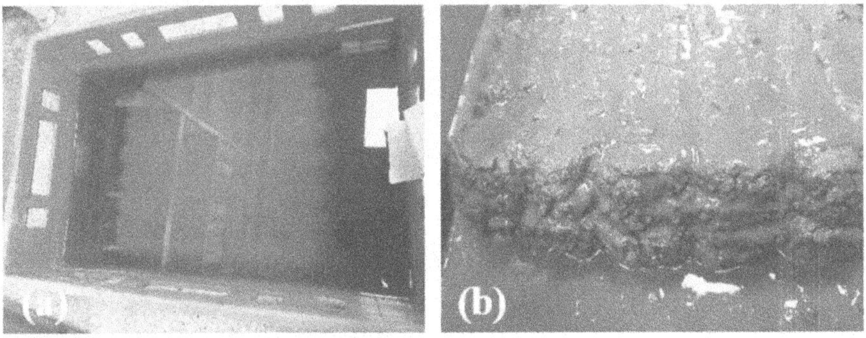

FIGURE 5.7 Photos from the project realisation: (a) the plate printed in the water and (b) the appearance of the plate after removing the water.

In this case, more adequate technology seems to be material extrusion because of its easier application in the underwater environment. Nowadays, this project is ongoing; some successful tests have been performed in laboratory conditions, and in the near future, trials on open water are planned.

5.5 CONCLUSIONS, CHALLENGES, AND FUTURE PROSPECTS

The main findings show that geopolymers could be effective materials for 3D printing technology for different applications in civil engineering as well as other areas. The usage of 3D printing can bring a number of benefits, including increasing work efficiency, enhancing safety, and reducing the impact on the environment. For that purpose, the process of material extrusion will be modified, and material properties will be improved. This chapter identifies the following areas as the most promising for the future development of geopolymers as a material for 3D printing technology:

One of the most promising areas for the application of 3D printed geopolymer materials is sustainable building that is based on a circular economy approach. The combination of usage by-products and waste in geopolymer materials strongly correlates with the minimum waste technology of 3D printing, giving an excellent synergy effect. This kind of solution should be focused on local sources of raw materials and be in line with environmental policy. Currently, the important issue is also the enhancement of mechanical properties of the 3D printed products. In this area, the investigation of different fibre reinforcements is necessary. The existing research defines the possible types of fibres to use, but further investigation for solving problems with technology is necessary. An interesting technological challenge is the technology of co-printed fibre combined with geopolymer paste.

The technology of 3D printed foamed geopolymers can have significant effects on isolation products. Despite the high potential, this research area is not so widely studied. The main challenges to be solved are connected with the stabilisation of the foaming process during additive manufacturing. 3D printing technology of geopolymers gives a huge perspective in the case of applications in harsh environments such as outside terrestrial applications and underwater projects. However, in this case, both elements – material and technology – are required to be developed for usage in in-situ conditions (relevant environment).

ACKNOWLEDGEMENTS

This research was supported by the project titled 'Development of geopolymer composites as a material for protection of hazardous wrecks and other critical underwater structures against corrosion' under the M-ERA.NET 3 program by the Polish National Centre for Research and Development, grant number M-ERA.NET3/2021/71/MAR-WRECK/2022, and co-financed with tax funds on the basis of the budget passed by the Saxon State Parliament (SAB).

REFERENCES

1. K. Korniejenko *et al.*, 'Additive manufacturing in underwater applications', *Applied Sciences*, vol. 14, no. 4, p. 1346, Feb. 2024, doi: 10.3390/app14041346.

2. Q.-P. Ma, H.-S. Nguyen, J. Hajnys, J. Mesicek, M. Pagac, and J. Petru, 'A bibliometric review on application of machine learning in additive manufacturing and practical justification', *Applied Materials Today*, vol. 40, p. 102371, Oct. 2024, doi: 10.1016/j.apmt.2024.102371.
3. H. G. Şahin and A. Mardani-Aghabaglou, 'Assessment of materials, design parameters and some properties of 3D printing concrete mixtures; a state-of-the-art review', *Construction and Building Materials*, vol. 316, p. 125865, Jan. 2022, doi: 10.1016/j.conbuildmat.2021.125865.
4. T. V. Sheeja, S. V. S. Jebadurai, D. Tensing, 'Additive manufacturing techniques in construction', *Research on Engineering Structures and Materials*, 2022, doi: 10.17515/resm2022.386ma1401.
5. R. Robayo-Salazar, R. Mejía De Gutiérrez, M. A. Villaquirán-Caicedo, and S. Delvasto Arjona, '3D printing with cementitious materials: Challenges and opportunities for the construction sector', *Automation in Construction*, vol. 146, p. 104693, Feb. 2023, doi: 10.1016/j.autcon.2022.104693.
6. R. S. Krishna *et al.*, 'Additive manufacturing of geopolymer composites for sustainable construction: Critical factors, advancements, challenges, and future directions', *Progress in Additive Manufacturing*, Aug. 2024, doi: 10.1007/s40964-024-00703-z.
7. P. Zhang, Y. Zheng, K. Wang, and J. Zhang, 'A review on properties of fresh and hardened geopolymer mortar', *Composites Part B: Engineering*, vol. 152, pp. 79–95, Nov. 2018, doi: 10.1016/j.compositesb.2018.06.031.
8. Y. Peng and C. Unluer, 'Development of alternative cementitious binders for 3D printing applications: A critical review of progress, advantages and challenges', *Composites Part B: Engineering*, vol. 252, p. 110492, Mar. 2023, doi: 10.1016/j.compositesb.2022.110492.
9. A. Baigarina, E. Shehab, and Md. H. Ali, 'Construction 3D printing: A critical review and future research directions', *Progress in Additive Manufacturing*, Feb. 2023, doi: 10.1007/s40964-023-00409-8.
10. P. Barve, A. Bahrami, and S. Shah, 'A comprehensive review on effects of material composition, mix design, and mixing regimes on rheology of 3D-printed geopolymer concrete', *TOBCTJ*, vol. 18, no. 1, p. e18748368292859, Jul. 2024, doi: 10.2174/0118748368292859240313061706.
11. Y. Chen *et al.*, 'Utilisation of municipal solid waste incineration fly ash as construction materials based on geopolymerization', *Resources, Conservation & Recycling Advances*, vol. 19, p. 200162, Nov. 2023, doi: 10.1016/j.rcradv.2023.200162.
12. K. K. D. Sungkono, I. Satyarno, H. Priyosulistyo, and I. Perdana, 'Corrosion resistance of high calcium fly ash based reinforced geopolymer concrete in marine environment', *Civil Engineering and Architecture*, vol. 11, no. 5A, pp. 3175–3189, Sep. 2023, doi: 10.13189/cea.2023.110827.
13. A. Yousaf, A. Al Rashid, and M. Koç, '3D printing of alkali-activated geopolymers for sustainable and circular economy advancements', *Circular Economy*, vol. 3, no. 3, p. 100101, Sep. 2024, doi: 10.1016/j.cec.2024.100101.
14. Y. Wu *et al.*, 'Geopolymer, green alkali activated cementitious material: Synthesis, applications and challenges', *Construction and Building Materials*, vol. 224, pp. 930–949, Nov. 2019, doi: 10.1016/j.conbuildmat.2019.07.112.
15. S. Singh, S. Mehla, S. K. Bhargava, and S. Ramakrishna, 'History and Evolution of Additive Manufacturing', in *Additive Manufacturing for Chemical Sciences and Engineering*, S. K. Bhargava, S. Ramakrishna, M. Brandt, and Pr. Selvakannan, Eds., Singapore: Springer Nature Singapore, 2022, pp. 19–51. doi: 10.1007/978-981-19-2293-0_2.
16. Y. W. Adugna, A. D. Akessa, and H. G. Lemu, 'Overview study on challenges of additive manufacturing for a healthcare application', *IOP Conference Series: Materials Science and Engineering*, vol. 1201, no. 1, p. 012041, Nov. 2021, doi: 10.1088/1757-899X/1201/1/012041.

17. J. Izdebska-Podsiadly, *Polymers for 3D Printing Methods, Properties, and Characteristics*, Kidlington: William Andrew, 2022.
18. S. A. M. Tofail, E. P. Koumoulos, A. Bandyopadhyay, S. Bose, L. O'Donoghue, and C. Charitidis, 'Additive manufacturing: Scientific and technological challenges, market uptake and opportunities', *Materials Today*, vol. 21, no. 1, pp. 22–37, Jan. 2018, doi: 10.1016/j.mattod.2017.07.001.
19. S. Molazadeh, F. Diba, and A. Hosseini, 'Anisotropic modeling of material behavior for additively manufactured parts made by material extrusion', *International Journal of Advanced Manufacturing Technology*, vol. 129, no. 7–8, pp. 3453–3473, Dec. 2023, doi: 10.1007/s00170-023-12508-5.
20. E. C. Balta, D. M. Tilbury, and K. Barton, 'Layer-to-layer stability of linear layer-wise spatially varying systems: Applications in fused deposition modeling', *IEEE Transactions on Control Systems Technology*, vol. 29, no. 6, pp. 2517–2532, Nov. 2021, doi: 10.1109/TCST.2020.3044237.
21. J. Frketic, T. Dickens, and S. Ramakrishnan, 'Automated manufacturing and processing of fiber-reinforced polymer (FRP) composites: An additive review of contemporary and modern techniques for advanced materials manufacturing', *Additive Manufacturing*, vol. 14, pp. 69–86, Mar. 2017, doi: 10.1016/j.addma.2017.01.003.
22. X. Peng, L. Kong, J. Y. H. Fuh, and H. Wang, 'A review of post-processing technologies in additive manufacturing', *JMMP*, vol. 5, no. 2, p. 38, Apr. 2021, doi: 10.3390/jmmp5020038.
23. K. A. Shiyas and R. Ramanujam, 'A review on post processing techniques of additively manufactured metal parts for improving the material properties', *Materials Today: Proceedings*, vol. 46, pp. 1429–1436, 2021, doi: 10.1016/j.matpr.2021.03.016.
24. A. Levy *et al.*, 'Ultrasonic additive manufacturing of steel: Method, post-processing treatments and properties', *Journal of Materials Processing Technology*, vol. 256, pp. 183–189, Jun. 2018, doi: 10.1016/j.jmatprotec.2018.02.001.
25. C. Oztan, R. Welch, and S. LeBlanc, 'Additive manufacturing of bulk thermoelectric architectures: A review', *Energies*, vol. 15, no. 9, p. 3121, Apr. 2022, doi: 10.3390/en15093121.
26. C. R. Gagg, 'Cement and concrete as an engineering material: An historic appraisal and case study analysis', *Engineering Failure Analysis*, vol. 40, pp. 114–140, May 2014, doi: 10.1016/j.engfailanal.2014.02.004.
27. N. B. Singh and B. Middendorf, 'Geopolymers as an alternative to Portland cement: An overview', *Construction and Building Materials*, vol. 237, p. 117455, Mar. 2020, doi: 10.1016/j.conbuildmat.2019.117455.
28. Y. H. M. Amran, R. Alyousef, H. Alabduljabbar, and M. El-Zeadani, 'Clean production and properties of geopolymer concrete; a review', *Journal of Cleaner Production*, vol. 251, p. 119679, Apr. 2020, doi: 10.1016/j.jclepro.2019.119679.
29. S. Riahi, A. Nemati, A. R. Khodabandeh, and S. Baghshahi, 'The effect of mixing molar ratios and sand particles on microstructure and mechanical properties of metakaolin-based geopolymers', *Materials Chemistry and Physics*, vol. 240, p. 122223, Jan. 2020, doi: 10.1016/j.matchemphys.2019.122223.
30. A. Tukaziban, C.-S. Shon, D. Zhang, J. R. Kim, J.-H. Kim, and C.-W. Chung, 'Synthesis and evaluation of geopolymer mixtures containing chronologically aged basic oxygen furnace slags', *Sustainability*, vol. 15, no. 24, p. 16934, Dec. 2023, doi: 10.3390/su152416934.
31. H. U. Ahmed *et al.*, 'Compressive strength of sustainable geopolymer concrete composites: A state-of-the-art review', *Sustainability*, vol. 13, no. 24, p. 13502, Dec. 2021, doi: 10.3390/su132413502.
32. M. S. Reddy, P. Dinakar, and B. H. Rao, 'A review of the influence of source material's oxide composition on the compressive strength of geopolymer concrete', *Microporous and Mesoporous Materials*, vol. 234, pp. 12–23, Nov. 2016, doi: 10.1016/j.micromeso.2016.07.005.

33. R. R. Bellum, R. Nerella, S. R. C. Madduru, and C. S. R. Indukuri, 'Mix design and mechanical properties of fly ash and GGBFS-synthesized alkali-activated concrete (AAC)', *Infrastructures*, vol. 4, no. 2, p. 20, May 2019, doi: 10.3390/infrastructures4020020.
34. A. M. M. A. Bakri, H. Kamarudin, M. Binhussain, I. K. Nizar, A. R. Rafiza, and Y. Zarina, 'Comparison of geopolymer fly ash and ordinary Portland cement to the strength of concrete', *Advanced Science Letters*, vol. 19, no. 12, pp. 3592–3595, Dec. 2013, doi: 10.1166/asl.2013.5187.
35. M. Kaya, F. Koksal, O. Gencel, M. J. Munir, and S. M. S. Kazmi, 'Influence of micro Fe_2O_3 and MgO on the physical and mechanical properties of the zeolite and kaolin based geopolymer mortar', *Journal of Building Engineering*, vol. 52, p. 104443, Jul. 2022, doi: 10.1016/j.jobe.2022.104443.
36. K. Ramujee and M. Potharaju, 'Abrasion resistance of geopolymer composites', *Procedia Materials Science*, vol. 6, pp. 1961–1966, 2014, doi: 10.1016/j.mspro.2014.07.230.
37. A. Fernández-Jiménez, A. Palomo, and C. López-Hombrados, 'Engineering properties of alkali-activated fly ash concrete', *Materials Journal*, vol. 103, no. 2, 2006, doi: 10.14359/15261.
38. S. Sbahieh, G. McKay, and S. G. Al-Ghamdi, 'Comprehensive analysis of geopolymer materials: Properties, environmental impacts, and applications', *Materials*, vol. 16, no. 23, p. 7363, Nov. 2023, doi: 10.3390/ma16237363.
39. M. A. M. Ariffin, M. A. R. Bhutta, M. W. Hussin, M. Mohd Tahir, and N. Aziah, 'Sulfuric acid resistance of blended ash geopolymer concrete', *Construction and Building Materials*, vol. 43, pp. 80–86, Jun. 2013, doi: 10.1016/j.conbuildmat.2013.01.018.
40. S. Sreenivasan and B. Kandasubramanian, 'Novel geomaterials for the remediation of toxic pollutants: A review', *Hybrid Advances*, vol. 3, p. 100057, Aug. 2023, doi: 10.1016/j.hybadv.2023.100057.
41. J. Jiang *et al.*, 'Synthesis of tailing slurry-based geopolymers for the highly efficient immobilisation of heavy metals: Behavior and mechanism', *Applied Clay Science*, vol. 247, p. 107199, Jan. 2024, doi: 10.1016/j.clay.2023.107199.
42. N. Ranjbar, M. Mehrali, M. Mehrali, U. J. Alengaram, and M. Z. Jumaat, 'Graphene nanoplatelet-fly ash based geopolymer composites', *Cement and Concrete Research*, vol. 76, pp. 222–231, Oct. 2015, doi: 10.1016/j.cemconres.2015.06.003.
43. P. Cong and Y. Cheng, 'Advances in geopolymer materials: A comprehensive review', *Journal of Traffic and Transportation Engineering (English Edition)*, vol. 8, no. 3, pp. 283–314, Jun. 2021, doi: 10.1016/j.jtte.2021.03.004.
44. W. M. Kriven *et al.*, 'Why geopolymers and alkali-activated materials are key components of a sustainable world: A perspective contribution', *Journal of American Ceramic Society*, vol. 107, no. 8, pp. 5159–5177, Aug. 2024, doi: 10.1111/jace.19828.
45. G. Habert, J. B. d'Espinose De Lacaillerie, and N. Roussel, 'An environmental evaluation of geopolymer based concrete production: Reviewing current research trends', *Journal of Cleaner Production*, vol. 19, no. 11, pp. 1229–1238, Jul. 2011, doi: 10.1016/j.jclepro.2011.03.012.
46. A. Akbarnezhad, M. Huan, S. Mesgari, and A. Castel, 'Recycling of geopolymer concrete', *Construction and Building Materials*, vol. 101, pp. 152–158, Dec. 2015, doi: 10.1016/j.conbuildmat.2015.10.037.
47. S. Kumar and R. Kumar, 'Geopolymer: Cement for low carbon economy', *The Indian Concrete Journal*, vol. 88, no. 7, pp. 29–37, Jul. 2014.
48. H. Zhong and M. Zhang, '3D printing geopolymers: A review', *Cement and Concrete Composites*, vol. 128, p. 104455, Apr. 2022, doi: 10.1016/j.cemconcomp.2022.104455.
49. P. Barve, A. Bahrami, and S. Shah, 'Geopolymer 3D printing: A comprehensive review on rheological and structural performance assessment, printing process parameters, and microstructure', *Frontiers in Materials*, vol. 10, p. 1241869, Nov. 2023, doi: 10.3389/fmats.2023.1241869.

50. K. Chen, Q. Liu, B. Chen, S. Zhang, L. Ferrara, and W. Li, 'Effect of raw materials on the performance of 3D printing geopolymer: A review', *Journal of Building Engineering*, vol. 84, p. 108501, May 2024, doi: 10.1016/j.jobe.2024.108501.
51. D.-W. Zhang, D. Wang, X.-Q. Lin, and T. Zhang, 'The study of the structure rebuilding and yield stress of 3D printing geopolymer pastes', *Construction and Building Materials*, vol. 184, pp. 575–580, Sep. 2018, doi: 10.1016/j.conbuildmat.2018.06.233.
52. G. Fang, W. K. Ho, W. Tu, and M. Zhang, 'Workability and mechanical properties of alkali-activated fly ash-slag concrete cured at ambient temperature', *Construction and Building Materials*, vol. 172, pp. 476–487, May 2018, doi: 10.1016/j.conbuildmat.2018.04.008.
53. A. S. Albidah, 'Effect of partial replacement of geopolymer binder materials on the fresh and mechanical properties: A review', *Ceramics International*, vol. 47, no. 11, pp. 14923–14943, Jun. 2021, doi: 10.1016/j.ceramint.2021.02.127.
54. S. Muthukrishnan, S. Ramakrishnan, and J. Sanjayan, 'Technologies for improving buildability in 3D concrete printing', *Cement and Concrete Composites*, vol. 122, p. 104144, Sep. 2021, doi: 10.1016/j.cemconcomp.2021.104144.
55. L. Ricciotti, A. Apicella, V. Perrotta, and R. Aversa, 'Geopolymer materials for extrusion-based 3D-printing: A review', *Polymers*, vol. 15, no. 24, p. 4688, Dec. 2023, doi: 10.3390/polym15244688.
56. B. Meskhi *et al.*, 'Analytical review of geopolymer concrete: Retrospective and current issues', *Materials*, vol. 16, no. 10, p. 3792, May 2023, doi: 10.3390/ma16103792.
57. P. Shakor, S. H. Chu, A. Puzatova, and E. Dini, 'Review of binder jetting 3D printing in the construction industry', *Progress in Additive Manufacturing*, vol. 7, no. 4, pp. 643–669, Aug. 2022, doi: 10.1007/s40964-021-00252-9.
58. M. Xia and J. Sanjayan, 'Method of formulating geopolymer for 3D printing for construction applications', *Materials & Design*, vol. 110, pp. 382–390, Nov. 2016, doi: 10.1016/j.matdes.2016.07.136.
59. S. Luhar and I. Luhar, 'Additive manufacturing in the geopolymer construction technology: A review', *The Open Construction & Building Technology Journal*, vol. 14, no. 1, pp. 150–161, Jul. 2020, doi: 10.2174/1874836802014010150.
60. M. Bagarić, Ed., *Energy Efficient Building Design and Legislation*, Paris: RILEM Publications, 2019.
61. H. S. Gökçe, M. Tuyan, and M. L. Nehdi, 'Alkali-activated and geopolymer materials developed using innovative manufacturing techniques: A critical review', *Construction and Building Materials*, vol. 303, p. 124483, Oct. 2021, doi: 10.1016/j.conbuildmat.2021.124483.
62. M. H. Raza, R. Y. Zhong, and M. Khan, 'Recent advances and productivity analysis of 3D printed geopolymers', *Additive Manufacturing*, vol. 52, p. 102685, Apr. 2022, doi: 10.1016/j.addma.2022.102685.
63. S. C. Paul, G. P. A. G. Van Zijl, M. J. Tan, and I. Gibson, 'A review of 3D concrete printing systems and materials properties: Current status and future research prospects', *Rapid Prototyping Journal*, vol. 24, no. 4, pp. 784–798, May 2018, doi: 10.1108/RPJ-09-2016-0154.
64. M. J. Bassan De Moraes, E. Y. Nagata, A. J. Felício Peres Duran, and J. A. Rossignolo, 'Alkali activated materials applied in 3D printing construction: A review', *Heliyon*, vol. 10, no. 5, p. e26696, Mar. 2024, doi: 10.1016/j.heliyon.2024.e26696.
65. M. H. Raza and R. Y. Zhong, 'A sustainable roadmap for additive manufacturing using geopolymers in construction industry', *Resources, Conservation and Recycling*, vol. 186, p. 106592, Nov. 2022, doi: 10.1016/j.resconrec.2022.106592.
66. L. Imtiaz, S. K. U. Rehman, S. Ali Memon, M. Khizar Khan, and M. Faisal Javed, 'A review of recent developments and advances in eco-friendly geopolymer concrete', *Applied Sciences*, vol. 10, no. 21, p. 7838, Nov. 2020, doi: 10.3390/app10217838.

67. J. Zhao *et al.*, 'Eco-friendly geopolymer materials: A review of performance improvement, potential application and sustainability assessment', *Journal of Cleaner Production*, vol. 307, p. 127085, Jul. 2021, doi: 10.1016/j.jclepro.2021.127085.
68. B. Panda, C. Unluer, and M. J. Tan, 'Extrusion and rheology characterisation of geopolymer nanocomposites used in 3D printing', *Composites Part B: Engineering*, vol. 176, p. 107290, Nov. 2019, doi: 10.1016/j.compositesb.2019.107290.
69. Z. Li, Y. Li, B. Shi, D. Tang, Y. Wang, and L. Hao, 'Dual gradient direct ink writing of functional geopolymer-based carbonyl-iron/graphene composites for adjustable broadband microwave absorption', *Ceramics International*, vol. 48, no. 7, pp. 9277–9285, Apr. 2022, doi: 10.1016/j.ceramint.2021.12.114.
70. H. Alghamdi and N. Neithalath, 'Synthesis and characterisation of 3D-printable geopolymeric foams for thermally efficient building envelope materials', *Cement and Concrete Composites*, vol. 104, p. 103377, Nov. 2019, doi: 10.1016/j.cemconcomp.2019.103377.
71. B. Nematollahi, S. H. Bong, M. Xia, and J. Sanjayan, 'Digital Fabrication of "Just-Add-Water" Geopolymers: Effects of Curing Condition and Print-Time Interval', in *Second RILEM International Conference on Concrete and Digital Fabrication*, vol. 28, F. P. Bos, S. S. Lucas, R. J. M. Wolfs, and T. A. M. Salet, Eds., in RILEM Bookseries, vol. 28, Cham: Springer International Publishing, 2020, pp. 93–102. doi: 10.1007/978-3-030-49916-7_10.
72. D. Mierzwiński, M. Łach, S. Gądek, W.-T. Lin, D. H. Tran, and K. Korniejenko, 'A brief overview of the use of additive manufacturing of con-create materials in construction', *Acta Innovations*, no. 48, pp. 22–37, Apr. 2023, doi: 10.32933/ActaInnovations.48.2.
73. G. Lazorenko and A. Kasprzhitskii, 'Geopolymer additive manufacturing: A review', *Additive Manufacturing*, vol. 55, p. 102782, Jul. 2022, doi: 10.1016/j.addma.2022.102782.
74. S. Qaidi, A. Yahia, B. A. Tayeh, H. Unis, R. Faraj, and A. Mohammed, '3D printed geopolymer composites: A review', *Materials Today Sustainability*, vol. 20, p. 100240, Dec. 2022, doi: 10.1016/j.mtsust.2022.100240.
75. F. A. Shilar, S. V. Ganachari, V. B. Patil, B. E. Bhojaraja, T. M. Yunus Khan, and N. Almakayeel, 'A review of 3D printing of geopolymer composites for structural and functional applications', *Construction and Building Materials*, vol. 400, p. 132869, Oct. 2023, doi: 10.1016/j.conbuildmat.2023.132869.
76. A. I. Yoris-Nobile *et al.*, 'Artificial reefs built by 3D printing: Systematisation in the design, material selection and fabrication', *Construction and Building Materials*, vol. 362, p. 129766, Jan. 2023, doi: 10.1016/j.conbuildmat.2022.129766.
77. O. Ly *et al.*, 'Optimisation of 3D printed concrete for artificial reefs: Biofouling and mechanical analysis', *Construction and Building Materials*, vol. 272, p. 121649, Feb. 2021, doi: 10.1016/j.conbuildmat.2020.121649.
78. P. Kinnunen, M. Karhu, E. Yli-Rantala, P. Kivikytö-Reponen, and J. Mäkinen, 'A review of circular economy strategies for mine tailings', *Cleaner Engineering and Technology*, vol. 8, p. 100499, Jun. 2022, doi: 10.1016/j.clet.2022.100499.
79. B. Figiela and K. Korniejenko, 'The possibility of using waste materials as raw materials for the production of geopolymers', *Acta Innovations*, no. 36, pp. 48–56, Sep. 2020, doi: 10.32933/ActaInnovations.36.4.
80. K. Korniejenko *et al.*, 'Mechanical and fracture properties of long fiber reinforced geopolymer composites', *Materials*, vol. 14, no. 18, p. 5183, Sep. 2021, doi: 10.3390/ma14185183.
81. K. Korniejenko *et al.*, 'Fracture behavior of long fiber reinforced geopolymer composites at different operating temperatures', *Materials*, vol. 15, no. 2, p. 482, Jan. 2022, doi: 10.3390/ma15020482.
82. J. Marczyk *et al.*, 'Properties of 3D printed concrete–geopolymer hybrids reinforced with aramid roving', *Materials*, vol. 15, no. 17, p. 6132, Sep. 2022, doi: 10.3390/ma15176132.

83. J. Marczyk *et al.*, 'Hybrid materials based on fly ash, metakaolin, and cement for 3D printing', *Materials*, vol. 14, no. 22, p. 6874, Nov. 2021, doi: 10.3390/ma14226874.
84. C. Ziejewska *et al.*, '3D printing of concrete-geopolymer hybrids', *Materials*, vol. 15, no. 8, p. 2819, Apr. 2022, doi: 10.3390/ma15082819.
85. K. Korniejenko, P. Kejzlar, and P. Louda, 'The influence of the material structure on the mechanical properties of geopolymer composites reinforced with short fibers obtained with additive technologies', *IJMS*, vol. 23, no. 4, p. 2023, Feb. 2022, doi: 10.3390/ijms23042023.
86. M. Sitarz *et al.*, 'Mechanical response of geopolymer foams to heating—Managing coal gangue in fire-resistant materials technology', *Energies*, vol. 15, no. 9, p. 3363, May 2022, doi: 10.3390/en15093363.
87. K. Korniejenko, K. Pławecka, and B. Kozub, 'An overview for modern energy-efficient solutions for lunar and martian habitats made based on geopolymers composites and 3D printing technology', *Energies*, vol. 15, no. 24, p. 9322, Dec. 2022, doi: 10.3390/en15249322.
88. M. Rudziewicz, M. Maroszek, K. Setlak, M. Góra, and M. Hebda, 'Optimisation of foams—Polypropylene fiber-reinforced concrete mixtures dedicated for 3D printing', *Materials*, vol. 17, no. 16, p. 4106, Aug. 2024, doi: 10.3390/ma17164106.
89. M. Sparkes, 'How deep is the Mariana trench?' [Online]. Available: https://www.newscientist.com/question/deep-mariana-trench/
90. H. Li *et al.*, 'Geopolymer composites for marine application: Structural properties and durability', *Cement and Concrete Composites*, vol. 152, p. 105647, Sep. 2024, doi: 10.1016/j.cemconcomp.2024.105647.
91. N. F. Shahedan *et al.*, 'Potential of fly ash geopolymer concrete as repairing and retrofitting solutions for marine infrastructure: A review', *Case Studies in Construction Materials*, vol. 20, p. e03214, Jul. 2024, doi: 10.1016/j.cscm.2024.e03214.

6 Industrial By-Products and Waste Materials for the Geopolymers in Construction

Ir. U. Johnson Alengaram

6.1 INTRODUCTION

In an era of heightened environmental awareness and advancing construction technologies, the search for materials that balance performance, sustainability, and cost-effectiveness has never been more critical. Traditional construction materials, particularly Portland cement, have long underpinned the built environment, forming the backbone of infrastructure worldwide [1]. However, the environmental toll of conventional cement production, accounting for roughly 8% of global CO_2 emissions driven by the high temperatures required in its manufacture and the extensive use of non-renewable resources [2], has sparked an imperative search for more sustainable alternatives. Amidst this backdrop, geopolymers have emerged as a transformative solution, promising a paradigm shift in construction materials.

Geopolymers are categorized as advanced inorganic polymers produced by the interaction of alumino-silicate minerals with alkaline reagents [3]. Unlike Portland cement, which is produced by limestone calcination at high temperatures, geopolymers are made at much lower temperatures using industrial and agro-based by-products, also known as supplementary cementitious materials (SCM), such as fly ash (FA), ground granulated blast-furnace slag (GGBS), silica fume (SF), rice husk ash (RHA), metakaolin (MK), palm oil fuel ash (POFA), palm oil clinker powder (POCP), coal bottom ash (CBA), etc. [4]. This not only reduces their carbon footprint but also capitalizes on materials that would otherwise contribute to waste, thus offering a sustainable advantage from both production and disposal perspectives [5]. As a result, there is a great interest in developing alternative binding materials that can reduce the carbon footprint and embodied energy of the product (concrete) while maintaining essential technical properties. Figure 6.1 depicts the use of industrial by-products and wastes responsibly.

DOI: 10.1201/9781003535683-6

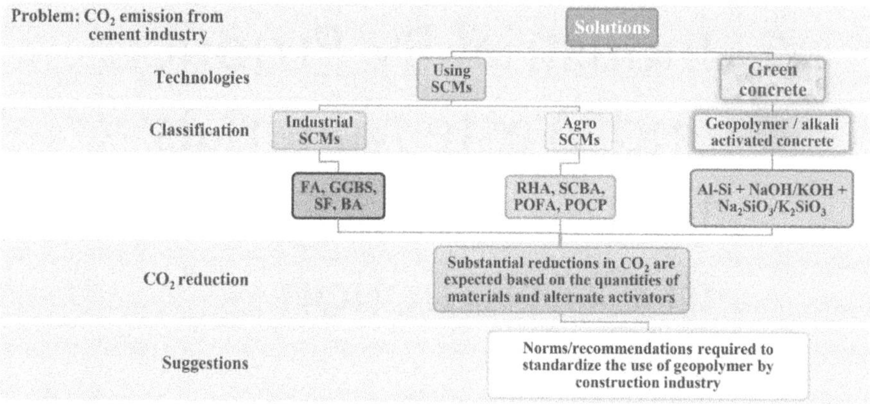

FIGURE 6.1 Utilization of industrial waste and by-products sustainably.

The properties of geopolymer materials distinguish them from traditional alternatives. Geopolymer concrete (GPC), for instance, exhibits exceptional durability, outstripping conventional concrete in resisting chemical attacks, such as chloride ingress, sulphate, and acid attack [3]. This makes it particularly suitable for harsh environments, including wastewater treatment facilities and marine structures. Furthermore, geopolymers exhibit remarkable fire resistance, withstanding high temperatures better than traditional concrete and thus providing enhanced safety in fire-prone scenarios [2].

Beyond concrete, the versatility of geopolymers extends to a range of construction materials including mortars, plasters, bricks, and tiles. Geopolymer mortars and plasters offer high adhesion strength and low shrinkage, making them ideal for both new constructions and repair applications [1]. Geopolymer bricks and blocks provide a sustainable alternative to conventional clay or concrete bricks, delivering comparable strength while enhancing thermal insulation. Similarly, geopolymer tiles and paving materials are increasingly utilized for their durability and aesthetic flexibility, suitable for both high-traffic and decorative applications [6].

This chapter explores the diverse applications of geopolymers within the construction industry, emphasizing their potential to revolutionize various aspects of building and infrastructure. It examines how geopolymers are utilized in structural components, including beams, columns, and pavements, and highlights their effectiveness in repair and rehabilitation projects. Additionally, the chapter delves into specialized applications such as fire-resistant panels and environmental protection measures, underscoring the adaptability and multifunctionality of geopolymer materials.

In the face of growing environmental regulations and the need for more resilient structures, understanding the potential of geopolymers becomes increasingly crucial. This chapter offers practical applications and continuous improvements that promise to strengthen its position in sustainable building. By shedding light on the transformative impact of geopolymers, this discussion not only showcases their current capabilities but also envisions their future contributions to a more sustainable and innovative built environment.

6.2 GEOPOLYMER

Geopolymer is an emerging technology that promotes a greener environment by using abundant industrial and agricultural by-products and waste. These materials serve as a complete replacement for cement, combined with chemical additives, fine and coarse aggregates, and water to form GPC [4].

Joseph Davidovits was the first to introduce the word ‘geopolymer’ in 1978 to describe a wide spectrum of materials composed of networks, chains, or links of inorganic monomer units or molecules. It displays a collection of mineral binders with an amorphous morphology and a chemical makeup similar to zeolites, known as ‘soil cement’ [7]. Geopolymers, unlike OPC, do not generate calcisilicate hydrates (C-S-Hs) to form the matrix and strength; instead, they use polycondensation of Si-Al to provide structural strength. Geopolymers are made up of two main ingredients: Si-Al source materials and alkaline or acidic activator solutions. The raw materials used as source materials must be fortified with aluminium (Al) and silicon (Si). The by-products and waste materials, such as GGBS, FA, SF, POFA, CBA, MK, POCP, and RHA, are reported to be precursors or source materials, as depicted in Figure 6.1 [8].

Geopolymer can be produced by activating alumina and silicate-rich raw material with alkali. Aqueous alkaline reagents can include sodium hydroxide (NaOH) or potassium hydroxide (KOH), as well as soluble silicates like sodium or potassium silicates (Na_2SiO_3 or K_2SiO_3). The chemical reaction between the source material and the alkaline reagent is often a polymerization process. The mechanism is hence referred to as geopolymerization [9].

Geopolymerization occurs in an alkaline medium when the alumino-silicates react with alkaline reagent(s) to produce monomer units of Si-O-Al-O links and combine to form a 3D polymeric chain-like structure. The 3D structures formed are of poly(sialate) (-Si-O-Al-O-), poly(sialate-siloxo) (-Si-O-Al-O-Si-O-), poly(sialate-disiloxo) (-Si-O-Al-O-Si-O-Si-O-), and sialate link, as illustrated in Figure 6.2. Sialate is an acronym for silicon-oxo-aluminate. A sialate network is composed of mainly aluminate (AlO_4) and silicate (SiO_4) molecules connected tetrahedrally and shares an oxygen (O) atom. Positive ions are needed in the framework cavities to counteract the negative charge of Al ions during fourfold coordination.

The empirical formula [10] of poly(sialates) is as follows:

$$Mn\,[-\,(SiO_2)\,z - AlO_2]\,n,\ wH_2O$$

where

M – cation (Na^+, K^+, Ca^{2+})

n – degree of polycondensation

z – 1, 2, 3

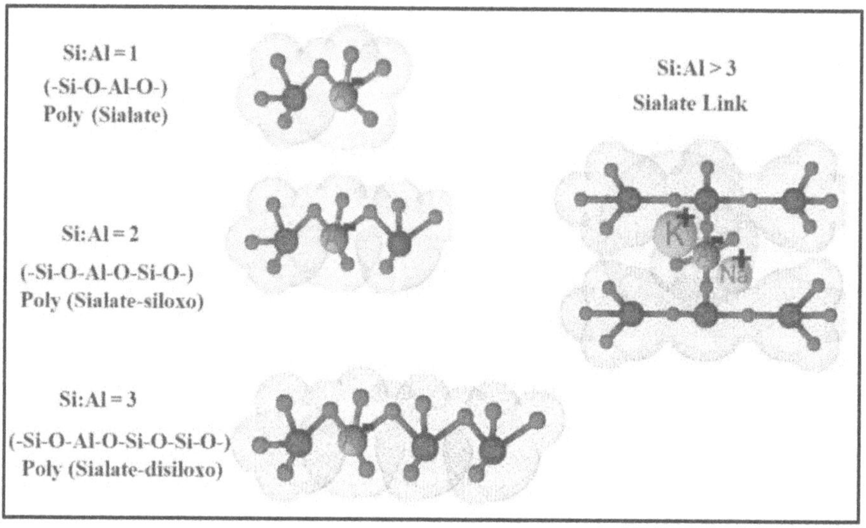

FIGURE 6.2 3D configuration of poly(sialate) [1].

6.3 MECHANISM OF GEOPOLYMER

The mechanism behind geopolymer formation involves several key steps:

i. Breakage or dissolution of bonds of silica-aluminate source materials in an alkaline medium (hydroxides of Na or K), releasing SiO_4 and AlO_4 tetrahedral units.
ii. The relatively unstable hydroxyl group in alkali allows these base units to interact and unite, resulting in T-O-T bonds (T = Al or Si).
iii. Finally, they rearrange and condense into geopolymer gels or other hydration products, enhancing mechanical and durability qualities [11], as shown in Figure 6.3.

Further, the geopolymer mechanism is explained in simple steps:

1. Source material: Geopolymers are typically made from Al- and/or Si-rich materials such as GGBS, FA, or natural minerals like kaolinite. These materials provide the necessary alumino-silicate framework.
2. Activation: The source material is mixed with an alkaline reagent. Common reagents include a combination of NaOH, commonly known as caustic soda or sodium hydroxide, and Na_2SiO_3, commonly known as water glass or sodium silicate in aqueous and other chemical reagents used, a combination of K_2SiO_3 and KOH. The alkaline solution breaks down the alumino-silicate structure into soluble silicate and aluminate species.

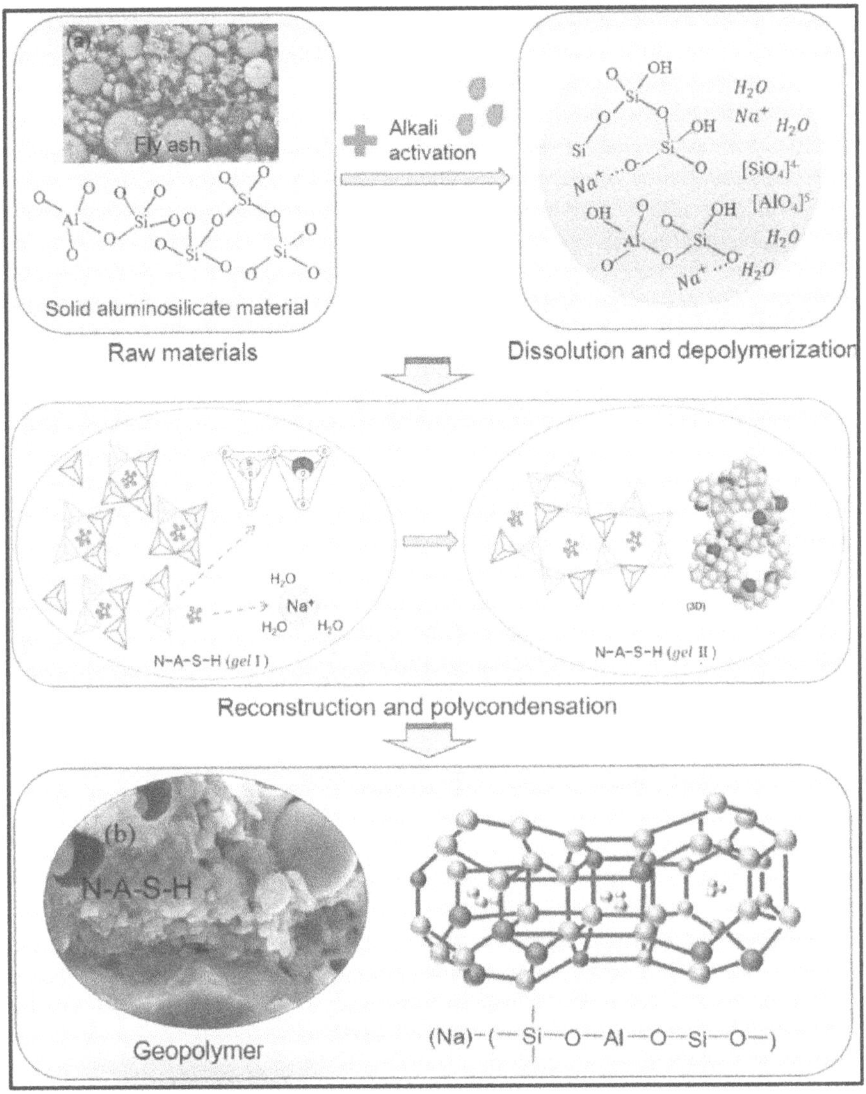

FIGURE 6.3 Mechanism of geopolymer [11].

3. Dissolution and polymerization: The activation process dissolves the alumino-silicate material into its constituent ions. These ions then react with each other, forming a network of alumino-silicate chains through a process called polycondensation. The primary reactions involve the formation of aluminum-oxygen-silicon (Al-O-Si) bonds.
   - Dissolution: The alkaline reagents break down the alumino-silicate material, releasing Al-Si ions into the solution.

- Polycondensation: The dissolved Si and Al ions subsequently react with one another to create a 3D polymeric network. This network is characterized by corner-sharing tetrahedra of Si and Al, with the Al generally in tetrahedral coordination with oxygen.

4. Gel formation: During polymerization, a gel-like structure forms. This gel is essentially the geopolymer matrix and contains the alumino-silicate network. It gradually hardens and solidifies over time, resulting in a stable and durable material.
5. Curing and hardening: The geopolymer gel continues to solidify and strengthen over time. Curing conditions, such as ambient temperature and humidity (steam), can affect the rate and extent of hardening. Geopolymers typically cure at room temperature; however, higher temperatures can speed up the process.

Their mechanism of action revolves around the formation of a robust alumino-silicate network that gives them their unique properties [12], as shown in Figure 6.4.

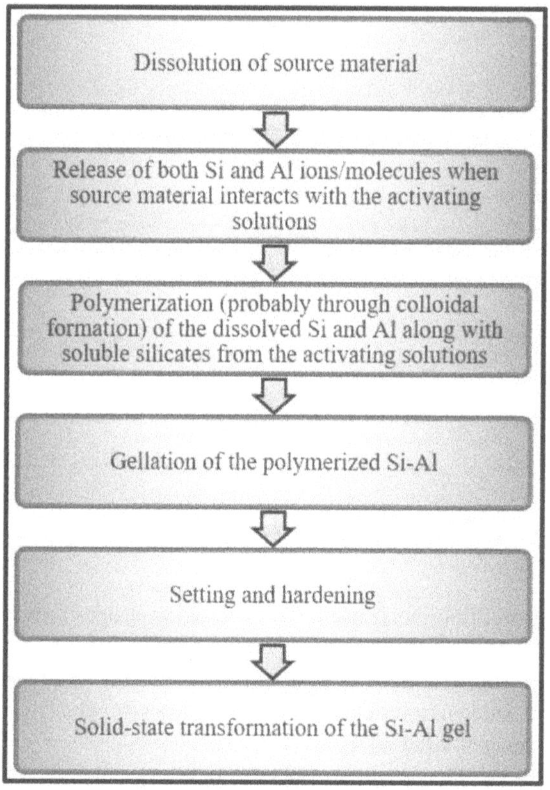

FIGURE 6.4 Geopolymer mechanism in simpler form.

6.4 MATERIALS USED FOR THE PREPARATION OF GEOPOLYMER

6.4.1 Precursors

Materials such as GGBS, FA, CBA, POFA, and POCP that are shown in Figure 6.5 and others have a reasonable amount of alumina and silica content, making them appropriate for substitution as Si-Al source materials for the creation of geopolymer instead of cement [10, 13]. The material description is presented in Table 6.1. The primary oxides found in the various material samples were largely silicon oxide and aluminium oxide, as shown in Table 6.2. The physical properties of various precursors are tabulated in Table 6.3. These chemical oxides are crucial because they may significantly boost the stability, endurance, and mechanical properties of geopolymers due to their large amount of reactive silica and alumina [3]. These precursors might be single or multi-material with a high Si and Al content, though. The strength of a geopolymer is governed by the precursors used. Calcium silicate facilitates the dissolution of alumino-silicate from the precursor structure; moreover, it integrates into the geopolymer matrix, serving as counterbalancing cations. Sodium aluminate increases the compressive strength by twice as much as non-sodium aluminate [2].

6.4.2 Chemical Reagents

The alkali activation of alumino-silicate is responsible for the creation of amorphous alumino-silicate gel, which has good mechanical and chemical characteristics [19]. Alkali activation in geopolymers is commonly achieved by the use of mixtures of Na_2SiO_3 and NaOH or K_2SiO_3 and KOH. Silicate anions have the most important role in improving geopolymerization in geopolymer materials through alkali activation. KOH has a higher alkalinity than NaOH and a lower activation potential. Higher NaOH concentrations enhanced the solubility

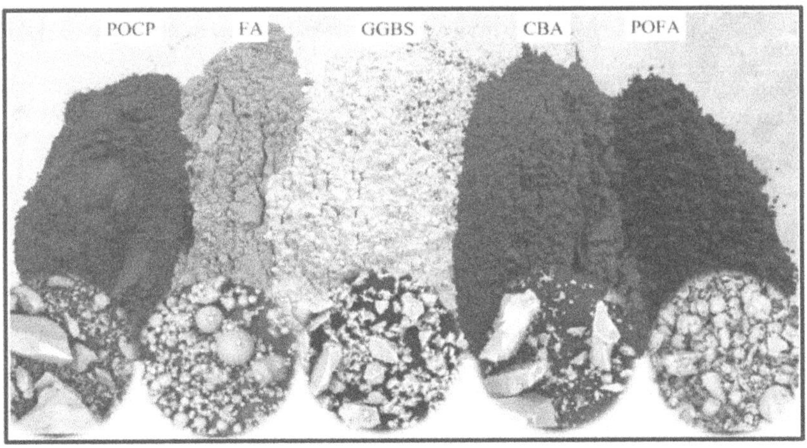

FIGURE 6.5 Geopolymer binders and SEM images of different precursors [13].

TABLE 6.1
Material Description

Materials	Description	Ref.
Fly ash (FA)	FA is a powdery waste or by-product produced by pulverized coal combustion in thermal power plants to generate electricity. It is a fine material and primarily consists of Al_2O_3, SiO_2, and Fe_2O_3	[2]
Groundgranulated blast-furnace slag (GGBS)	GGBS is a residual by-product or waste produced by the steel industry and is generated by rapid cooling of molten slag collected from blast furnace using water or steam, further grinding it into fine powder. It consists of CaO, SiO_2, Al_2O_3, and MgO	[14]
Coal bottom ash (CBA)	CBA is a granular, coarse waste or by-product from coal combustion in power plants. It collects at the bottom of the furnace and can be used as lightweight aggregates. Further, the large chunks can be grinded to get the powdery material that can be used as SCMs. CBA typically consists of unburned carbon, Al_2O_3, SiO_2, and various minerals	[13]
Eco-processed pozzolan (EPP)	EPP is a residual by-product produced by calcining spent bleaching earth (SBE), a solid waste material resulting from the waste residues of refined crude oil from the palm oil industry. It primarily consists of Al_2O_3, SiO_2, Fe_2O_3 and CaO	[5, 15]
Palm oil fuel ash (POFA)	POFA is a waste by-product generated from palm oil biomass burning, primarily in power plants and mills. It is composed of fine particles rich in SiO_2 and Al_2O_3 which gives it pozzolanic properties	[16, 17]
Palm oil clinker powder (POCP)	POCP is a residual by-product derived from the incineration of palm oil biomass and is primarily composed of residues from palm kernel shells and other palm oil waste materials. After being processed, it turns into a fine powder. It exhibits valuable pozzolanic properties due to the presence of Al_2O_3 and SiO_2	[16]
Rice husk ash (RHA)	RHA is a by-product or waste from the rice industry, produced by burning rice husks. It is lightweight, highly porous, and rich in silica, making it an effective pozzolanic material as it consists of around 85–95% amorphous silica along with small amounts of alumina and other minerals	[10]
Metakaolin (MK)	MK is a calcined clay material derived from kaolin, a type of clay. It is produced by heating kaolin to high temperatures (around 600–800°C), which transforms its structure and enhances its pozzolanic properties. MK is composed of silica and alumina and has a high reactivity due to its amorphous nature	[2]

TABLE 6.2
Oxides Composition of the Precursors for Geopolymer

	Oxides Composition (%)								
Materials	SiO_2	Al_2O_3	CaO	MgO	Fe_2O_3	Na_2O	K_2O	LOI	Ref.
FA	57.20	29.50	0.21	0.90	5.91	0.61	0.91	3.61	[15]
GGBS	34.60	17.90	36.50	6.48	0.46	0.62	0.39	0.07	[14]
CBA	50.41	15.52	2.53	0.90	6.96	0.39	0.75	5.73	[13]
EPP	54.93	9.42	9.28	5.81	9.28	0.34	1.41	2.07	[15]
POFA	50.49	0.66	3.95	3.46	1.95	0.08	5.56	15.48	[16]
TPOFA	59.24	0.84	4.05	4.03	1.99	0.12	5.73	1.48	[16]
POCP	59.12	5.72	3.98	4.20	5.27	0.20	7.24	3.68	[16]
RHA	93.46	0.58	1.03	0.51	0.52	0.08	1.82	7.76	[10]
MK	53.37	42.21	0.16	–	0.31	–	0.07	1.89	[18]

rate of the source material by providing enough OH^- to dissolve alumina and silica [9].

However, NaOH/KOH concentrations over 14 M (molarity) have been shown to result in a loss in mechanical strength due to the instability of oligomeric silicate, which shifts towards mono-silicate, delaying the geopolymerization process [20–22]. Furthermore, the rate of geopolymerization reaction is boosted in the presence of Na_2SiO_3/K_2SiO_3, which increases the dissolution of alumina and silica [9]. The geopolymer mortar/concrete preparation is shown in Figure 6.6.

TABLE 6.3
Physical Properties of the Different Precursors for Geopolymer

	Physical Properties					
Materials	Specific Gravity	Colour	Shape	Specific Surface Area (m^2/kg)	Median Particle Size (μm)	Ref.
FA	2.17	Grey	Spherical	350	12.35	[15]
GGBS	2.90	White	Irregular	512	16.61	[16]
CBA	2.76	Dark grey	Irregular	432	48.05	[16]
EPP	2.44	Brown	Irregular	223	25.69	[15]
POFA	2.10	Dark grey	spherical and porous	509	19.41	[16]
TPOFA	2.13	Light grey	Irregular	523	19.19	[16]
POCP	2.53	Dark grey	Irregular	424	47.85	[16]
RHA	2.45	Dark grey	Irregular	–	<45.00	[10]
MK	2.51	Off-white	Angular	–	<45.00	[18]

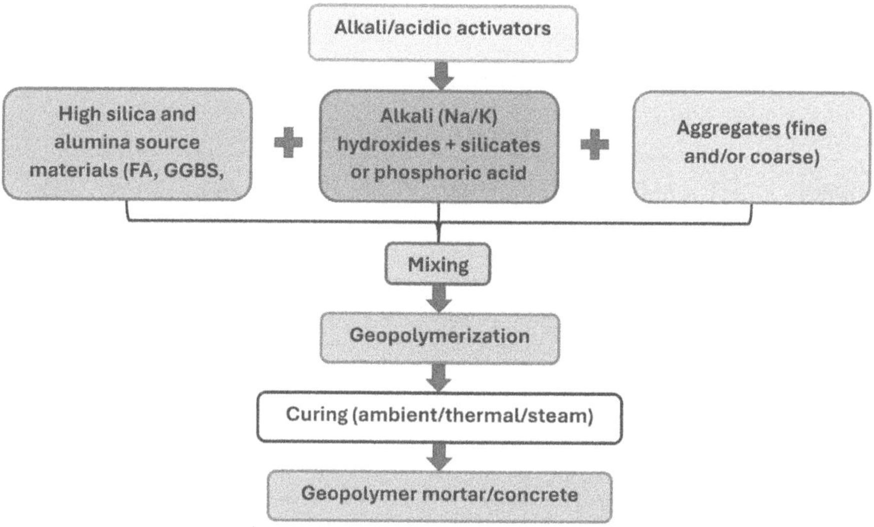

FIGURE 6.6 Geopolymer production process.

6.5 MECHANICAL PROPERTIES OF GEOPOLYMER

GPC is an innovative and sustainable construction material that offers several mechanical properties, often superior to traditional Portland cement concrete. The following are some key mechanical properties of GPC:

1. Compressive strength: GPC generally demonstrates a high compressive strength, frequently surpassing 20 MPa and reaching 100 MPa or more, depending on the formulation and curing circumstances [1, 4].
2. Splitting tensile strength: As known for OPC-based concrete, the splitting tensile strength of GPC could be in the range of 8–10% of its compressive strength; however, it can range between 2 and 7 MPa. Some formulations and additions can improve this characteristic up to 10 MPa [15, 23].
3. Flexural strength: Flexural strength usually ranges from 5 to 10 MPa. This property can also be improved with the addition of fibres or specific aggregate types and can go up to 20 MPa [15, 23].
4. Elastic modulus: The elastic modulus or modulus of elasticity (MoE) for GPC is similar to or slightly greater than that of conventional concrete, often between 20 and 30 GPa, depending on the mix design [15, 23].
5. Bond strength: It typically shows good bond strength with reinforcement materials and gives a bond strength between 10 and 30 MPa with a maximum slip of 0.4 to 1.3 mm with different diameters of reinforcing bars (12 to 16 mm), contributing to the overall structural integrity [24].
6. Compression toughness: The compressive toughness of GPC is higher than that of conventional cement concrete, making it suitable for heavy stable structures [20].

TABLE 6.4
Mechanical Properties of GPC

S. No.	Mechanical Properties	Range
1	Compressive strength	20–100 MPa and more **[1, 4]**
2	Splitting tensile strength	2–10 MPa **[15, 23]**
3	Flexural strength	5–20 MPa **[15, 23]**
4	Modulus of elasticity	20–30 GPa **[15, 23]**
5	Bond strength	10–30 MPa (0.4 to 1.3 mm slip) **[24]**

7. Shrinkage and creep: GPC generally has lower shrinkage and creep compared to traditional concrete, making it suitable for large structures where dimensional stability is critical [25].

These properties can be modified by a variety of parameters, including the type of precursor materials (e.g., FA, slag), the alkaline reagents employed, fibres used, curing conditions, and mix design. Properly built GPC can therefore serve as a long-term and high-performance alternative to ordinary concrete. The mechanical parameters of GPC are reported in Table 6.4. The flexure and split-tensile strength can further be enhanced by adding fibres in the GPC to get fibre-reinforced GPC [26].

6.6 APPLICATIONS OF GEOPOLYMER

GPC might be used to make precast railway sleepers, bridge components, and other precast construction elements. It is beneficial in harsh maritime settings as well as soils heavy in carbon or sulphate. Similarly, in a more acidic environment, geopolymer demonstrates higher resistance to acidic nature and may be suitable for applications such as mining, some industrial sectors, and sewage systems [27]. GPC can be used in the construction of rigid pavement owing to its high stiffness and flexural and tensile strength [28]. Geopolymers have a range of innovative applications in the construction industry, leveraging their unique properties to improve the performance and sustainability of building materials [29]. The following are some specific applications of geopolymers in construction:

Geopolymer Concrete

- Structural concrete: Used in constructing buildings, bridges, and other infrastructure. GPC offers high mechanical strength, durability, and chemical resistance, making it suitable for demanding environments.
- High-performance concrete: Ideal for applications requiring high mechanical properties and durability, such as high-rise buildings and heavy-load structures.
- Precast elements: Ideal for precast concrete elements such as panels, blocks, and modular units. Its fast setting and curing times make it suitable for prefabricated construction [3].

Geopolymer Mortar

- Repair and rehabilitation: Used for repairing and restoring damaged concrete structures. It adheres well to existing concrete and provides excellent resistance to environmental degradation [30].
- Tile adhesives: Geopolymer mortars are used as adhesives for ceramic tiles and other finishing materials due to their strong bonding properties.

Geopolymer Bricks and Blocks

- Load-bearing bricks: Used in walls and partitions, offering high strength and thermal insulation. Geopolymer bricks are often produced from industrial by-products, making them a sustainable option.
- Insulating blocks: Designed to provide thermal insulation in addition to structural support, improving energy efficiency in buildings [6, 19].

Geopolymer Panels and Slabs

- Facade panels: Used for building facades due to their durability and ability to resist weathering. Geopolymer panels can be customized to meet aesthetic and functional requirements.
- Flooring slabs: Utilized for floor systems, offering durability and strength suitable for both residential and commercial applications [31].

Geopolymer Coatings

- Protective coatings: Applied to steel reinforcement surfaces to enhance resistance to corrosion and environmental degradation. Commonly used in industrial facilities and infrastructure exposed to harsh conditions [32].

Geopolymer Insulation

- Thermal insulation panels: Used in walls, roofs, and floors to improve energy efficiency. Geopolymer-based insulation materials can offer better thermal performance than conventional options [19].

Geopolymer Lightweight Concrete

- Lightweight blocks: Used for non-load-bearing applications or as infill materials in construction. This type of concrete reduces the overall weight of structures, which can lower transportation and handling costs [33].

Geopolymer Paving

- Paving stones and tiles: Used for exterior paving applications such as sidewalks, plazas, and driveways. Geopolymer paving materials offer excellent resistance to environmental factors and wear [6].

Geopolymer Fireproofing

- Fire-resistant: Provides high fire resistance, making it suitable for applications requiring stringent fire safety standards.
- Fireproof panels: Used in areas requiring high fire resistance, such as in building facades and fire barriers. Geopolymer materials maintain their structural integrity and provide insulation under extreme heat conditions [27].

FIGURE 6.7 (a) Three-Storey Queensland's University GCI Building using structural GPC and (b) precast FA/slag-based GPC floor panel parts [26].

Overall, the usage of geopolymer materials in construction is motivated by their environmental benefits, such as the utilization of industrial and agro wastes and/or by-products and lower CO_2 emissions as compared to ordinary cement. Their strength, durability, and adaptability make them a popular choice for a variety of building projects, as illustrated below.

6.6.1 Global Change Institute (GCI) Constructed Using Geopolymer Concrete at Queensland University, Australia

In Australia, the production and application of GPC and precast products have grown in recent years. The construction of the world's first public building using structural GPC in the University of Queensland's GCI, designed by HASSELL in partnership with Bligh Tanner and Wagner, is a testament to geopolymer's eco-friendly concrete application that successfully employed FA/slag-based GPC for structural purposes [31] as presented in Figure 6.7a and b.

6.6.2 Brisbane West Wellcamp Airport (BWWA), Australia

The BWWA is Australia's first greenfield public airport, as shown in Figure 6.8. BWWA began operating commercial flights in November 2014. The BWWA's apron,

FIGURE 6.8 Brisbane West Wellcamp Airport (BWWA), Australia [29].

FIGURE 6.9 Geopolymer concrete house (first cement-free concrete house in Malaysia in Universiti Malaya campus for bus drivers).

turning node, and taxiway pavement construction is one of the large-scale cast-in-situ structures built with paving-grade GPC. This project represents a significant engineering achievement as the 'world's largest geo-polymer concrete project'. BWWA was constructed with over 40,000 m^3 (100,000 tonnes) of geopolymer concrete, making it the largest deployment of this new form of concrete in the world [34].

6.6.3 First Geopolymer Concrete House in Malaysia at Universiti Malaya, Kuala Lumpur, Malaysia for Bus Drivers

The first cement-free GPC house for bus drivers was constructed by using embryonic construction materials at Universiti Malaya, Malaysia, as shown in Figure 6.9. The use of FA, GGBS, and POFA as whole cement replacement enabled the construction of the house without conventional cement. Further, the use of M-sand (fine aggregates), POC (palm oil clinker), and SA (slag aggregates) as coarse aggregates proved how local materials could be effectively utilized. The use of industrialized building systems (IBS) in the form of interlocking panels and blocks was a further testament to bringing the materials to fruition in the development and demonstration of GPC by utilizing locally available sustainable construction materials.

6.7 CONCLUSION

The use of geopolymers in construction materials represents a fundamental shift in building technology, driven by their excellent performance qualities and environmental benefits. Geopolymers, which are composed of alumino-silicate minerals activated by alkaline solutions, offer higher strength, durability, and resistance to harsh conditions, making them a potential alternative to typical Portland cement-based products. The diverse applications of geopolymers, ranging from structural concrete and repair mortars to bricks, blocks, and panels, demonstrate their versatility and efficacy. GPC stands out for its high compressive strength and reduced environmental impact, positioning it as a viable option for constructing robust structural elements while addressing sustainability goals. Geopolymer mortars facilitate effective repairs and serve as resilient adhesives, proving their value in maintaining and enhancing existing structures.

In masonry, geopolymer bricks and blocks not only provide load-bearing capabilities but also enhance thermal insulation, contributing to energy-efficient building practices. Geopolymer panels and slabs offer durability and aesthetic flexibility, making them suitable for prefabricated construction and façade applications. Additionally, geopolymer coatings extend the lifespan of surfaces by providing superior protection against environmental wear.

The use of geopolymer-based insulation materials further demonstrates their potential to improve energy efficiency while offering excellent fire resistance. Lightweight GPC and formwork systems enhance construction efficiency and sustainability by reducing overall structural weight and enabling reusable formwork solutions.

Overall, the application of geopolymers in construction materials represents a forward-thinking approach to modern building practices. Their ability to utilize industrial by-products further enhances their eco-friendliness, aligning with global efforts towards reducing carbon footprints. As the industry continues to explore and optimize geopolymer technologies, their role in transforming construction practices and contributing to sustainable development becomes increasingly prominent. The adoption of geopolymers not only solves performance and environmental problems but also sets the path for future innovative and durable construction materials.

REFERENCES

1. El Alouani M, Saufi H, Aouan B, Bassam R, Alehyen S, Rachdi Y, El Hadki H, El Hadki A, Mabrouki J, Belaaouad S, Hassan EZ. A comprehensive review of synthesis, characterization, and applications of aluminosilicate materials-based geopolymers. Environmental Advances. 2024 Mar 29: 100524. DOI: https://doi.org/10.1016/j.envadv.2024.100524
2. Shehata N, Sayed ET, Abdelkareem MA. Recent progress in environmentally friendly geopolymers: A review. Science of The Total Environment. 2021 Mar 25; 762: 143166. DOI: https://doi.org/10.1016/j.scitotenv.2020.143166
3. Madirisha MM, Dada OR, Ikotun BD. Chemical fundamentals of geopolymers in sustainable construction. Materials Today Sustainability. 2024 May 19: 100842. DOI: https://doi.org/10.1016/j.mtsust.2024.100842
4. Kanagaraj B, Anand N, Alengaram UJ, Praveen B, Tattukolla K. Performance evaluation on engineering properties and sustainability analysis of high strength geopolymer concrete. Journal of Building Engineering. 2022 Nov 15; 60: 105147. DOI: https://doi.org/10.1016/j.jobe.2022.105147
5. Ahmat AM, Alengaram UJ, Shamsudin MF, Alnahhal AM, Ibrahim MS, Ibrahim S, Rashid RS. Assessment of sustainable eco-processed pozzolan (EPP) from palm oil industry as a fly ash replacement in geopolymer concrete. Construction and Building Materials. 2023 Jul 17; 387: 131424. DOI: https://doi.org/10.1016/j.conbuildmat.2023.131424
6. Nawaz M, Heitor A, Sivakumar M. Geopolymers in construction-recent developments. Construction and Building Materials. 2020 Nov 10; 260: 120472. DOI: https://doi.org/10.1016/j.conbuildmat.2020.120472

7. Farooq F, Jin X, Javed MF, Akbar A, Shah MI, Aslam F, Alyousef R. Geopolymer concrete as sustainable material: A state of the art review. Construction and Building Materials. 2021 Nov 1; 306: 124762. DOI: https://doi.org/10.1016/j.conbuildmat.2021.124762
8. Sharmin A, Alengaram UJ, Jumaat MZ, Yusuf MO, Kabir SA, Bashar II. Influence of source materials and the role of oxide composition on the performance of ternary blended sustainable geopolymer mortar. Construction and Building Materials. 2017 Jul 30; 144: 608–623. DOI: https://doi.org/10.1016/j.conbuildmat.2017.03.178
9. Wong CL, Mo KH, Alengaram UJ, Yap SP. Mechanical strength and permeation properties of high calcium fly ash-based geopolymer containing recycled brick powder. Journal of Building Engineering. 2020 Nov 1; 32: 101655. DOI: https://doi.org/10.1016/j.jobe.2020.101655
10. Ng C, Alengaram UJ, Wong LS, Mo KH, Jumaat MZ, Ramesh S. A review on microstructural study and compressive strength of geopolymer mortar, paste and concrete. Construction and Building Materials. 2018 Oct 20; 186: 550–576. DOI: https://doi.org/10.1016/j.conbuildmat.2018.07.075
11. Yuan L, Wang H, He T, Gao S. Review on the effect collision between hazardous metal ions and geopolymer as adsorbents or in situ stabilization/solidification. Applied Clay Science. 2024 Mar 1; 249: 107258. DOI: https://doi.org/10.1016/j.clay.2024.107258
12. Ranjbar N, Mehrali M, Behnia A, Alengaram UJ, Jumaat MZ. Compressive strength and microstructural analysis of fly ash/palm oil fuel ash based geopolymer mortar. Materials & Design. 2014 Jul 1; 59: 532–539. DOI: https://doi.org/10.1016/j.matdes.2014.03.037
13. Alnahhal AM, Alengaram UJ, Ibrahim MS, Yusoff S, Metselaar HS, Johnson PG. Synthesis of ternary binders and sand-binder ratio on the mechanical and microstructural properties of geopolymer foamed concrete. Construction and Building Materials. 2022 Sep 26; 349: 128682. DOI: https://doi.org/10.1016/j.conbuildmat.2022.128682
14. Tanu HM, Unnikrishnan S. Mechanical strength and microstructure of GGBS-SCBA based geopolymer concrete. Journal of Materials Research and Technology. 2023 May 1; 24: 7816–7831. DOI: https://doi.org/10.1016/j.jmrt.2023.05.051
15. Gaddafi AK, Alengaram UJ, Bunnori NM, Muhammad SI, Ibrahim S, Sumesh M. Mechanical properties, flexural behaviour, and ductility characteristics of fibre-reinforced geopolymer mortar. Construction and Building Materials. 2023 Nov 3; 403: 133109. DOI: https://doi.org/10.1016/j.conbuildmat.2023.133109
16. Sumesh M, Alengaram UJ, Jumaat MZ, Mo KH, Singh R, Nayaka RR, Srinivas K. Chemo-physico-mechanical characteristics of high-strength alkali-activated mortar containing non-traditional supplementary cementitious materials. Journal of Building Engineering. 2021 Dec 1; 44: 103368. DOI: https://doi.org/10.1016/j.jobe.2021.103368
17. Alnahhal AM, Alengaram UJ, Ibrahim MS, Radwan MK, Ayough P. Extraction of home brewed sodium silicate from palm oil fuel ash and its effect on alkali activated materials. Construction and Building Materials. 2023 Dec 1; 407: 133440. DOI: https://doi.org/10.1016/j.conbuildmat.2023.133440
18. Zhang HY, Liu JC, Wu B. Mechanical properties and reaction mechanism of one-part geopolymer mortars. Construction and Building Materials. 2021 Mar 1; 273: 121973. DOI: https://doi.org/10.1016/j.conbuildmat.2020.121973
19. Mo KH, Bong CS, Alengaram UJ, Jumaat MZ, Yap SP. Thermal conductivity, compressive and residual strength evaluation of polymer fibre-reinforced high volume palm oil fuel ash blended mortar. Construction and Building Materials. 2017 Jan 15; 130: 113–121. DOI: https://doi.org/10.1016/j.conbuildmat.2016.11.005
20. Bashar II, Alengaram UJ, Jumaat MZ. Enunciation of embryonic palm oil clinker based geopolymer concrete and its engineering properties. Construction and Building Materials. 2022 Feb 7; 318: 125975. DOI: https://doi.org/10.1016/j.conbuildmat.2021.125975

21. Darvish P, Alengaram UJ, Poh YS, Ibrahim S, Yusoff S. Performance evaluation of palm oil clinker sand as replacement for conventional sand in geopolymer mortar. Construction and Building Materials. 2020 Oct 20; 258: 120352. DOI: https://doi.org/10.1016/j.conbuildmat.2020.120352
22. Kabir SA, Alengaram UJ, Jumaat MZ, Yusoff S, Sharmin A, Bashar II. Performance evaluation and some durability characteristics of environmental friendly palm oil clinker based geopolymer concrete. Journal of Cleaner Production. 2017 Sep 10; 161: 477–492. DOI: https://doi.org/10.1016/j.jclepro.2017.05.002
23. Gaddafi AK, Alengaram UJ, Bunnori NM, Ibrahim MS, Ibrahim S, Govindasami S. Enhancement of ductility characteristics of fiber-reinforced ternary geopolymer mortar. Journal of Building Engineering. 2024 Apr 1; 82: 108141. DOI: https://doi.org/10.1016/j.jobe.2023.108141
24. Romanazzi V, Leone M, Aiello MA, Pecce MR. Bond behavior of geopolymer concrete with steel and GFRP bars. Composite Structures. 2022 Nov 15; 300: 116150. DOI: https://doi.org/10.1016/j.compstruct.2022.116150
25. Singh B, Ishwarya G, Gupta M, Bhattacharyya SK. Geopolymer concrete: A review of some recent developments. Construction and Building Materials. 2015 Jun 15; 85: 78–90. DOI: https://doi.org/10.1016/j.conbuildmat.2015.03.036
26. Bashar II, Alengaram UJ, Jumaat MZ, Islam A, Santhi H, Sharmin A. Engineering properties and fracture behaviour of high volume palm oil fuel ash based fibre reinforced geopolymer concrete. Construction and Building Materials. 2016 May 15; 111: 286–297. DOI: https://doi.org/10.1016/j.conbuildmat.2016.02.022
27. Almutairi AL, Tayeh BA, Adesina A, Isleem HF, Zeyad AM. Potential applications of geopolymer concrete in construction: A review. Case Studies in Construction Materials. 2021 Dec 1; 15: e00733. DOI: https://doi.org/10.1016/j.cscm.2021.e00733
28. Podolsky Z, Liu J, Dinh H, Doh JH, Guerrieri M, Fragomeni S. State of the art on the application of waste materials in geopolymer concrete. Case Studies in Construction Materials. 2021 Dec 1; 15: e00637. DOI: https://doi.org/10.1016/j.cscm.2021.e00637
29. Assi LN, Carter K, Deaver E, Ziehl P. Review of availability of source materials for geopolymer/sustainable concrete. Journal of Cleaner Production. 2020 Aug 1; 263: 121477. DOI: https://doi.org/10.1016/j.jclepro.2020.121477
30. Zailani WW, Apandi NM, Adesina A, Alengaram UJ, Faris MA, Tahir MF. Physico-mechanical properties of geopolymer mortars for repair applications: Impact of binder to sand ratio. Construction and Building Materials. 2024 Jan 19; 412: 134721. DOI: https://doi.org/10.1016/j.conbuildmat.2023.134721
31. Islam A, Alengaram UJ, Jumaat MZ, Bashar II, Kabir SA. Engineering properties and carbon footprint of ground granulated blast-furnace slag-palm oil fuel ash-based structural geopolymer concrete. Construction and building materials. 2015 Dec 30; 101: 503–521. DOI: https://doi.org/10.1016/j.conbuildmat.2015.10.026
32. Ranjbar N, Mehrali M, Mehrali M, Alengaram UJ, Jumaat MZ. High tensile strength fly ash based geopolymer composite using copper coated micro steel fiber. Construction and Building Materials. 2016 Jun 1; 112: 629–638. DOI: https://doi.org/10.1016/j.conbuildmat.2016.02.228
33. Kupaei RH, Alengaram UJ, Jumaat MZ, Nikraz H. Mix design for fly ash based oil palm shell geopolymer lightweight concrete. Construction and Building Materials. 2013 Jun 1; 43: 490–496. DOI: https://doi.org/10.1016/j.conbuildmat.2013.02.071
34. Glasby T, Day J, Genrich R, Aldred J. EFC geopolymer concrete aircraft pavements at Brisbane West Wellcamp Airport. Concrete. 2015 Aug; 2015: 1–9. https://www.geopolymer.org/wp-content/uploads/GP-AIRPORT.pdf

7 Geopolymers in Environment-Friendly Fire-Retardation Coatings for Metallic Structures

Kaibo Wang, Longyuan Li, and Huirong Le

7.1 INTRODUCTION

Coatings are commonly applied to the surface of an object or substrate to improve its surface properties, such as appearance, adhesion, abrasion resistance, corrosion resistance, roughness, scratch resistance, thermal resistance, and wettability [1]. It can also help to maintain or improve the object's mechanical and physical properties. In addition, coatings allow for an enhancement in the structure or substrate's service life, which is associated with ease of maintenance and a reduction in refurbishment costs [1, 2].

One of the significant applications of coating is to protect steel in the building and construction sectors. The steel structure is one mainstream building structural system due to the high strength-weight ratio and ductility, which enable the designing of fascinating slim and light structures [3]. In addition, steel structures allow for the reduction of construction costs. Although the steel is not susceptible to fire, it experiences significant strength and stiffness reduction at elevated temperatures [3]. At a temperature above 550°C, steel can lose up to 50% of its yield strength [4]. In an extreme fire event, an unprotected steel structure may collapse due to a loss in mechanical properties at extremely high temperatures [5]. Therefore, a major design consideration in high-rise buildings during a fire is to prevent steel from reaching critical temperature and prevent fire-induced structural collapse by providing a suitable passive fireproofing system. It gives sufficient time to evacuate humans and valuables [5].

The conventional passive fireproofing system typically includes passive fire protection materials (e.g., cement-based coatings and thermal insulation panels) or reactive materials (e.g., intumescent coatings) [6]. Intumescent coatings have been used as a passive fire protection material for over 20 years to prevent structural steel in high-rise buildings from collapsing [5]. They provide a high-quality finish, lightweight structure, and good fire ratings on buildings. However, the outdoor exposed coatings have poor corrosion resistance due to the presence of water-sensitive ingredients in the formulations [5]. Besides, the presence of organic binders, such

DOI: 10.1201/9781003535683-7

as epoxy resin and polyurethane, emit smoke from combustion and release odours during a fire event, which has a negative impact on the environment and human health. Cement-based coatings are more economical and accessible in application but have major limitations in their weight, thickness, and poor aesthetics [5]. Thus, innovations in conventional passive and novel coatings are constantly developing and growing rapidly.

In recent years, extensive research has been conducted to investigate the feasibility of geopolymers as a sustainable alternative to ordinary Portland cement (OPC) [7]. Geopolymer is a three-dimensional inorganic polymer that is derived from the reaction between an aluminosilicate source and alkaline/acidic activators [8, 9]. Geopolymer exhibits excellent thermal stability, relatively high strength, and fire and chemical resistance. More interestingly, geopolymer can be fabricated using industrial solid waste, such as fly ash (FA) and slug, and it produces less carbon dioxide (CO_2) as well as consumes less energy during manufacturing than traditional cement [10, 11]. It was reported that the estimated CO_2 emissions during the production of geopolymer concrete could be reduced by up to 45% than traditional cement concrete [11]. Apart from the environmental effect, the polymerisation reaction speed of geopolymer is rapid, and the three-dimensional network structure can be readily formed. Moreover, FA-based geopolymer can significantly reduce the leaching of toxic and harmful heavy metal ions, which has attracted more attention and research [12].

The geopolymer paste as a coating material could become an ideal solution to minimise deterioration by protecting the surface and increasing the chemical and thermal resistance of the substrate [13]. Despite the fact that a few geopolymer coating reviews have been conducted, a comprehensive review, including the current conventional coatings and novel geopolymer coating that is only focused on the metal substrate, as well as the comparison between them, is rarely reported. The chapter aims to summarise the current use of conventional passive coating and assess the potential use of novel geopolymer coating on the metal substrate, practically on the steel structure. The research progress of geopolymer coatings is investigated. Finally, a few investigations regarding the comparison between conventional coating and geopolymer coating are discussed.

7.2 THE CONVENTIONAL COATING ON METAL STRUCTURE

7.2.1 Passive Coatings

The conventional passive coatings are mainly based on Portland cement, vermiculite, gypsum, and other materials. During the on-site application, they are mixed in water with fillers and binders, which are then applied by spraying at a thickness of a few inches [5]. They provide fire protection by thermal insulation effects and water release from a few minutes to several hours. Cement-based materials exhibit low thermal conductivity and incombustibility and do not swell under heat, which is ideal for the application of fire-resistive materials [14].

Compared with traditional intumescent coatings, Portland cement-based coatings provide superior properties, including anti-ageing, anti-cracking, and anti-swelling, when exposed to water. In addition, the simple and flexible spraying method significantly shortens the installation time, thus reducing the construction and repair costs [3, 15]. Moreover, due to their inorganic nature, they will not produce smoke and release odours in the event of a fire. Therefore, the research and development of cement-based coatings have attracted much attention.

Despite the above-mentioned advantages, the drawback of the cement-based coating is also unavoidable. The mechanical properties of cement-based coating deteriorate significantly at high temperatures [14]. More seriously, spalling may occur, meaning a sudden and rapid brittle failure occurs. To overcome these deteriorations, inorganic porous fillers, such as perlite, hollow glass beads, aerogels, and expanded polystyrene, are typically used to improve the mechanical properties of cement-based materials. The presence of short fibres, such as polypropylene (PP) fibres and steel fibres, could enhance the spalling performance, ensuring the structural integrity of the coating during fire [14]. Moreover, adding calcined products, such as FA, blast furnace slag, and bottom ash, improves the thermal and mechanical properties of cement-based materials [14]. Ma et al. [14] summarised the advantages and disadvantages of various porous lightweight fillers, including FA cenospheres, hollow glass beads, aerogel, and phase change materials. It was stated that the filler type needs to be correctly chosen to satisfy the desired thermal and mechanical performance by considering the cost, mechanical strength loss, ease of mixing, and complexity of the encapsulation process.

The involvement of fibre can reduce the chance of spalling under high temperatures, which is influenced by fibre type, size, and weight. At present, PP fibres are the most widely used fibres to resist the spalling of cement-based materials at high temperatures effectively [16–18]. This is because PP fibres melt at relatively low temperatures (160–170°C). The melted PP fibres are then rapidly absorbed by the slurry, and finally, the remaining empty channels will be interconnected to form a network of cracks in the matrix [16]. In addition, the presence of PP fibres can effectively reduce the pore pressure and thermal strain of the concrete, significantly reducing the range of the stress field during heating [18]. Similarly, the fire spalling resistance could be improved by adding other types of polymer fibres with a low melting point, such as nylon and PVA fibre [14]. Although the involvement of polymer fibres can effectively resist concrete spalling, it cannot improve the residual compressive strength and elastic modulus of the concrete at high temperatures [16]. The use of steel fibres allows the prevention of spalling while improving the residual mechanical properties after heating to some extent, but it could only reduce the degree of spalling severity and duration of the severe spalling. Thus, the combination of polymer and steel fibres was investigated to improve both properties of cement-based coating [19]. It was stated that the PP fibres with a lower melting point would first melt and form a broader channel to release water vapour. As the temperature continues to rise, the steel fibres inside the matrix control the expansion of internal cracks during heating. As a result, both the spalling resistance and residual mechanical properties can be improved [19].

The use of cement-based materials in steel structural coatings exhibits excellent thermal insulation properties and spalling resistance performance, and mechanical properties are improved after elevated temperature. In addition, cement-based coatings are generally more economical and more accessible in application compared to intumescent coating. However, in an actual application, the required thickness of the cement-based coating is significantly larger than that of intumescent coating, which enormously affects the structural self-weight and aesthetics [14]. The inherent brittleness and poor bonding properties of cement-based coatings result in delamination to substrate and loss of fire protective performance, resulting in a deterioration of the fire resistance and structural stability. Currently, only a few publications related to the mechanism of fire protection, explosion suppression of cement-based coating, and interfacial bonding strength to metal can be found, which needs further investigation.

7.2.2 Reactive Materials

Intumescent coatings have been increasingly used as a fire protection material for both commercial and industrial applications due to their high-quality finish, lightweight, speed of construction, and fair fire ratings [20]. Intumescent coatings are thermally reactive materials, and nowadays they are usually composed of a combination of organic and inorganic components bound together in a polymer matrix [3, 21]. They are usually composed of an acid source, e.g., ammonium polyphosphate, a carbonaceous compound, e.g., polymers, a blowing agent, e.g., melamine, binders, and additives. The effect of these compositions on the coating performance has been explored extensively [3]. These types of coatings are insensitive to weather and water compared with alkali silicate-based coatings, which are commonly used for passive fire protection of steel-framed structures. In addition, they provide a finish that does not detract from the appearance of the exposed steelwork, as in the case of cementitious coatings. Therefore, they are increasingly used in modern airports, skyscrapers, sports complexes, shopping malls, and hotels [20].

Intumescent coating is reactive because it swells due to heat exposure, increasing its original thickness and decreasing in density many times, producing a porous media that acts as an insulating layer to protect the substrate (Figure 7.1) [3, 5, 22]. The intumescence process generally involves the following steps: a mineral acid is

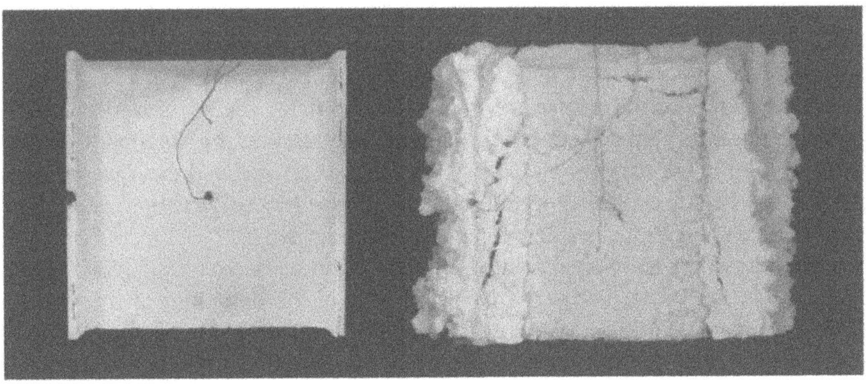

FIGURE 7.1 Steel structure protected with an intumescent coating before (left) and after (right) fire test. (Reprinted with permission from ref. [3]. Copyright 2019 Elsevier and [24].)

released by breaking the acid source, resulting in dehydration and carbonisation of char formers. The blowing agent then decomposes to release gases and is trapped in the melted matrix, making it swell. An insulating multi-cellular protective char layer is formed. This carbonaceous char shield limits the heat transfer from the source to the substrate and prevents further degradation of the underlying material [20, 23].

As mentioned previously, the main ingredients of intumescent are an acid source, a carbon source, and a blowing agent. The intumescent coating needs to be well designed and formulated to form an efficient protective char, as its structure significantly impacts the heat barrier properties. A coating should form a large char volume with a thick and continuous inner char structure, while the char should be compact, having a smaller cell size, narrower cell size distribution, and closed solid foam [5].

Ammonium polyphosphate is the most widely used acid source in intumescent formations, and it releases acid above 250°C in the event of a fire. However, the disadvantage is the water sensitivity and poor compatibility with the binder. Other salts, such as melamine phosphate and melamine pyrophosphate, are also widely used in intumescent formations [20]. Melamine phosphate has slightly less solubility in water and is a suitable choice in combination with ammonium polyphosphate. Moreover, melamine pyrophosphate is thermally more stable and also has less water solubility. It also exhibits better water resistance and gives char with better thermal barrier properties than melamine phosphate [5].

The presence of char former allows the formation of a carbonaceous char by dehydration in the presence of an acid source [5]. Pentaerythritol is a widely used charring agent in combination with ammonium polyphosphate as an acid source in intumescent formulations due to its low melting point [22]. However, the poor water solubility limits its use, especially for stable weather applications [5]. Starches, sugars, and cellulose that contain pendant hydroxyl groups are also attractive in intumescent formulations due to their low cost and sustainability [25].

Blowing agents, typically melamine, are commonly used in intumescent formulations, which decompose to form a considerable amount of gases that cause the carbon-rich mass to bubble and foam, producing a thick insulating layer [20]. The use of binders improves protective performance. They also have to degrade over the same temperature range as other intumescent fire-retardant additives in the formulation so as to produce intumescence, leading to a carbonaceous char. Various polymeric resins can be used as a binder in intumescent coatings [5]. The choice of binder depends upon the service environment to which the coating is exposed and its expected durability. Waterborne acrylics are mostly used in dry, internal locations, while solvent-borne acrylics are used in an internal or sheltered external location. Solvent-less or solvent-borne epoxy coatings are used in any location where high weatherability and protective performance are anticipated [5].

Inorganic fillers and fibres are also involved in intumescent formulations, which improve the fire protection performance as well as the durability of the coating. Fillers such as magnesium hydroxide are reported to increase bonding strength to the metal substrate due to its effective interface adhesion [26]. Inorganic compounds such as borides, nitrides, titanium, or other metals are chemically inert and exhibit a much higher thermal decomposition temperature. They could stabilise char in intumescent formulations [5]. The addition of fibres in the intumescent coating has been well investigated. It was reported that alumina and silica fibres increase the strength of the residual char, thus elongating the fireprotective performance [27]. High-temperature ceramic fibres could enhance the toughness of the char [28]. More recently, the addition of nano-scale additives, such as carbon nanotubes, in intumescent coatings has been widely explored as it was proved that they could enhance the fire protection performance of the coatings [5, 29].

Thin intumescent coatings are typically solvent- or water-based with a dry thickness ranging between 400 and 3000 μm. They are mainly applied in the built environment and used for cellulosic fire conditions. In contrast, thick intumescent coatings are usually epoxy-based (two-component systems) with thickness in the order of a few centimetres. Thick intumescent coatings are mainly applied in industrial applications, such as the oil and gas industry, and they are used for hydrocarbon fire conditions [3, 6]. Intumescent coatings usually consist of primer, base coat, and top coat. The primer ensures good adhesion between the metal substrate and the coating and offers improved corrosion resistance, while the top coat is usually for decorative or durability reasons when exposed to weathering and an aggressive environment [3, 20].

Intumescent coatings offer passive fire protection with a rating of a few minutes to several hours. They also provide an aesthetic appearance [6]. Moreover, they could be used in cellulosic fire and hydrocarbon fibre, including jet fire, which effectively prolongs fire-induced structural collapse. However, the application usually involves a long drying time, which is sometimes difficult to use on-site. In addition, the outdoor exposed coatings have poor corrosion resistance due to the presence of water-sensitive ingredients in the formulations. Moreover, the formation of carbonaceous char may exhibit poor cohesive strength and fragility, leading to the falling of the coatings during a fire event [5]. Finally, organic binders, such as epoxy resin and polyurethane, emit smoke from combustion and release odours during a fire event, which negatively impacts the environment and human health. Thus, continuous improvement of the intumescent coating is still needed.

7.3 THE POTENTIAL OF GEOPOLYMER COATING ON METAL STRUCTURE

In recent years, geopolymer coatings have attracted much attention due to their better workability, higher mechanical strength, excellent resistance to chemicals and heat, and easier handling. In addition, geopolymer coatings offer excellent durability and long-term performance. Moreover, the green and environmentally friendly characteristics make geopolymer an admirable coating material for many applications [30, 31].

The mixture compositions with regard to the activator solution and binder content, as well as the curing conditions, are the influencing factors for the mechanical and durability properties of geopolymer coatings [4, 31]. A defect-free and integral coating is essential to provide strong adhesion to the substrate. Therefore, it is necessary to balance the preparation process of the coating, including water evaporation, geopolymerisation, the formation of a bond between the coating and the substrate, and the techniques used to apply the coating [1].

The common techniques used to apply geopolymer coatings on substrate surfaces are dipping, spraying, blading, brushing, and extruding [32]. The dipping method is convenient and simple, but the major drawback is the inability to control the thickness of the applied layer accurately. Similar to the dipping method, the brushing method leaves relatively thick coatings on substrate surfaces. This may limit its use when layers of small thickness are required, for example, the possibility of mismatched properties between the coating and the substrate [1]. Geopolymer coatings can also be manually applied to the surface via a blading method. It usually has a high shrinkage rate in geopolymer paste, which consequently forms microcracks on the surface in atmospheric conditions [33]. The spray method is a low-cost and efficient method which can precisely control the thickness of the applied layer. In contrast, the disadvantage is that a flowable coating suspension is required [33, 34]. Nevertheless, the spraying method still offers the highest application potential for geopolymer coating.

As a coating material applied on metal surfaces in the building and construction industry, the coating performance, such as adhesion strength, fire resistance, corrosion resistance, and shrinkage, significantly impacts the safety and service life of existing structures.

7.3.1 Adhesion Strength

Adhesion strength is generally defined as the capability of coating materials to adhere to the substrates [31]. For a coating to be retained and to perform its function, adhesion to the substrate must tolerate mechanical stress, thermal stress, or elastoplastic distortions and the environment [35]. There are two types of bonding failures: the coating layer debonding from the interface between coatings and substrate and the substrate or the coating layer failure tears. The system is likely to fail at its weakest spots [36]. The adhesions of geopolymer coatings are typically influenced by the properties of the coating, the quality and state of the substrate, and the construction technique of coatings [37].

It was well proved that the effect of Si/Al ratio and Na/Al ratio significantly impacts the adhesive strength [32, 38]. Temuujin et al. [38] studied the effect of the Si/Al ratio on adhesion for FA-based geopolymer coating applied to mild steel via the dipping method. Results showed that the increase in the Si/Al ratio induced increases in the adhesive strength, with a maximum strength of 3.5 MPa at a Si/Al ratio of 3.5. Khan et al. [39] investigated the effect of the Na/Al ratio on the adhesive strength of FA-based coating applied on steel by dipping. The adhesive strength of the prepared samples tends to increase, followed by a reduction as the Na/Al ratio increases from 0.6 to 1.2.

The water content also plays a vital role in forming geopolymer coating and is correlated to the Si/Al and Na/Al ratios [40]. Water is consumed in the dissolution of the aluminosilicate solid by alkaline hydrolysis but released in condensation of the dissolved oligomers. As an aqueous medium, high water content is conducive to full hydration of the precursor materials, leading to soft geopolymer binders. This is beneficial for applying geopolymer as adhesive coatings [41, 42]. The geopolymerisation process is induced at low water content, resulting in a reduction of porosity in the

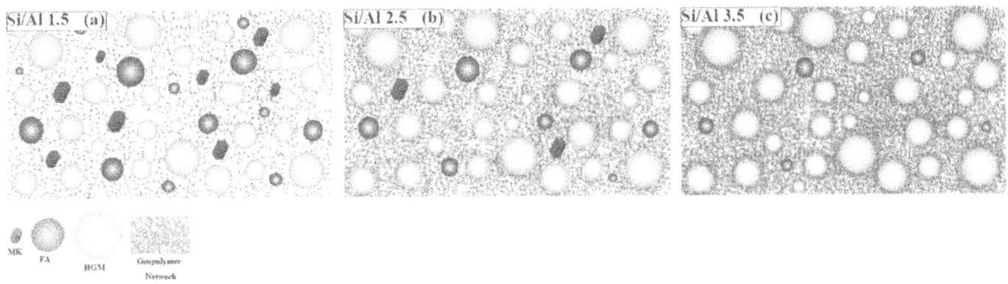

FIGURE 7.2 Schematic illustration of dissolution-reorganisation mechanism in geopolymer formation at varying Si/Al mole ratios: (a) 1.5, (b) 2.5, and (c) 3.5 in HGM/geopolymer composites. (Reprinted with permission from ref. [44]. Copyright 2022 Elsevier, and [24].)

binder. The workability of the binder decreased, making the coating hard to deposit [43]. It was suggested that the adhesion strength tends to decrease with increasing water content [8].

Tatlisu et al. [44] prepared FA-based geopolymer coating with the addition of hollow glass microspheres (HGM). The composite coating was applied on hot (300°C) metal surfaces via the spray deposition technique. The effect of Si/Al ratio, HGM content, and Na/Al ratio on adhesion strength and thermal conductivity was investigated. As expected, for the samples without HGM, the increase in Si/Al ratio (1.5 to 3.5) resulted in a significant increase in adhesion strength (1 to 6.5 MPa) due to the increasing dissolved geopolymer phase (Figure 7.2), while the thermal conductivity of the samples decreased slightly. When the HGM was added, the adhesion strength decreased slightly (5.5 to 5.0 MPa at a Si/Al ratio of 2.5) as the wt% of HGM increased from 0 to 10% with a fixed Si/Al ratio. However, as the amount of HGM increased further to 15%, the adhesion strength dropped significantly to 1.0 MPa. The thermal conductivity of the composite coating decreased significantly with the addition of the HGM. Therefore, depending on the application, there should be a trade-off between the adhesion strength and the thermal conductivity.

Tomar et al. [45] prepared geopolymer coating containing FA, red mud, metakaolin (MK), and blast furnace slag, and the amount of red mud on the adhesion strength was evaluated. The coating material was applied on mild steel via the spray coating technique, and the thickness of the coatings was in the range of 240–290 μm. Generally, the adhesion strength increased with the addition of red mud, and the strength of all samples was reported between 10.80 and 12.15 MPa. This may be due to the synergistic effect of red mud and FA constituents, which emerged as iron that contains inorganic phases.

The bonding strength of geopolymer coating to stainless steel and aluminium is generally weak due to only a physical bond being formed [8]. Chromium (Cr) present in stainless steel inhibited the geopolymerisation of the binder on a stainless-steel substrate and formed a weak bond [46]. Excessive reaction with aluminium substrates results in detrimental corrosion. This caused very weak bonding between the coating and substrates [47]. Temuujin [32] studied the adhesion strength of FA-based geopolymer coating on mild steel and stainless steel (Figure 7.3). Results showed that it exhibits higher adhesion strength on mild steel (2.7 MPa) compared with stainless

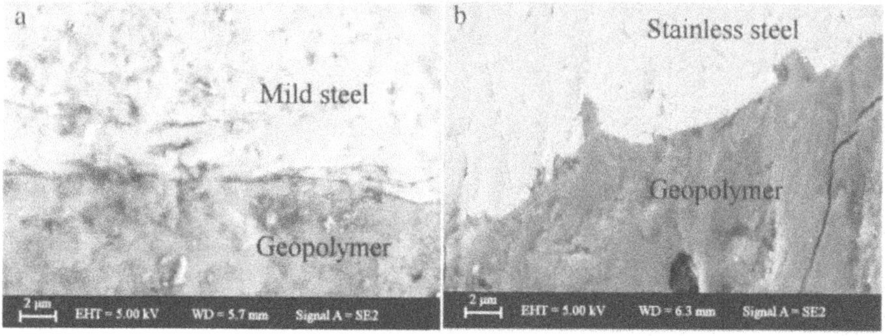

FIGURE 7.3 Interface between geopolymer coating and (a) mild steel and (b) stainless steel. (Reprinted with permission from ref. [32]. Copyright 2009 Elsevier and [24].)

steel (0.25 MPa). Yong [46] suggested that the growth of synthetic geopolymeric gel is more rapid when placed on an iron substrate due to chemical bonding, while the presence of Cr in stainless steel inhibits the growth of geopolymeric gel, and weak bonding was observed.

High adhesive strength likely involved mechanical interlocking or chemical bonds between the coating and the substrates, apart from the physical bond. This could be achieved by increasing the surface roughness of the substrates, leading to a higher contact area between the coating and the substrate [8, 48]. Mechanical treatment (sandblasting) and chemical treatment (nitrophosphoric acid and sanitation) could be carried out on steel and aluminium plates [40]. De Barros et al. [49] suggested that mechanical treatments were more effective than chemical treatments for geopolymer coating applied on steel and aluminium substrates. Wang et al. [50] examined the effect of substrate surface characteristics (relative area) on the adhesion properties of geopolymer coatings. The relative area was determined by area scale fractal analysis, which can effectively characterise the surface roughness. It was reported that the bond strength reduced with the decreasing relative area. However, Temuujin et al. [32] claimed that the difference in adhesion strength of various geopolymer compositions applied on metals was clearly independent of surface roughness. Even if the metal surface is treated the same way, the surface roughness of each substrate could vary due to the time and power of grinding. The relationship between substrate surface roughness and adhesion strength in the coating industry is widely investigated, which provides valuable guidance for geopolymer coating applied on the metal surface [40].

7.3.2 Fire Resistance

Steel, as a load-bearing structural component, is widely used in the building and construction industry, and fireproofing is one of the most critical properties. The yield stress of steel will decrease as the temperature rises. At a temperature above 550°C, steel can lose up to 50% of its yield strength [4]. Fire protection is rated for the time that passive fire protection can withstand a standard fire resistance test, such as a 2-h rating for steel beam protection. Due to their inorganic nature, geopolymer coatings exhibit higher strength, durability, and fire resistance characteristics compared to conventional coating materials [31, 36]. It was well indicated that the effects of composition, additives, and thickness of coatings have a significant influence on the thermal resistance of the geopolymer coating [8].

Temuujin et al. [51] conducted a direct flame test (flame temperature of 1100–1200°C) for MK-based geopolymer coating. Results indicated that the coating could withstand the direct flame for 30 min, which denoted high structural integrity. Temuujin et al. [38] also explored the microstructure of the geopolymer coating at elevated temperatures. It was observed that the material exhibited shrinkage with a temperature increase of up to 820°C. The metal substrates expanded after heating while the geopolymer shrunk, resulting in a loss of adhesion. The formation of cracks during heating due to heat flow directly to the metal substrate caused a loss of strength. Bakharev [52] suggested that KOH-activated FA-based geopolymer coating had better thermal stability than NaOH-activated geopolymer coating. The fire resistance property was also improved with increasing K_2O and MK content.

Nicoară et al. [53] prepared a geopolymer coating containing waste glass powder, FA, NaOH, KOH, and borax decahydrate (chemical reagents). The coatings were applied on a metal surface via spray technique and tested in direct contact with a propane flame. An uncoated steel plate was used as a reference, and the flame test showed that the back face temperature rose rapidly, reaching the critical value of 500°C in the first 2 min. This temperature was critical for the structural strength of steel. For the samples containing glass powder, FA, and borax (N_C_0.58 and N7_K3_C_0.58), the back face temperature remained broadly stable at below 500°C for more than 1 h, indicating an excellent thermal resistance property. During the flame test, no exfoliation of the coating layer was observed, indicating excellent adhesion between the coating and the substrate. It is also worth mentioning that the addition of glass powder increased the temperature at which intumescence occurred, which was beneficial to the fire resistance.

Sarazin et al. [54] explored the influence of coating thickness on the fire resistance test by adding a foaming agent and surfactant (CetylTrimethyl Ammonium Bromide, or CTAB). The temperature profiles for three geopolymer coatings were compared during the burn-through tests. Whatever the coating is considered, the rise in temperature is lower than for the uncoated steel plates. The maximum decrease in temperature (–251°C) was obtained for geopolymer coating modified with both a foaming agent and surfactant, whereas the lowest (–122°C) was obtained for the reference coating (i.e., without a foaming agent and surfactant) compared to uncoated steel plate. This means that the foaming structure provided a better thermal barrier than bulk geopolymer, and the addition of foam stabiliser (CTAB) enhanced the thermal barrier effect as well. This is expected as both additives increase the foam thickness.

Bhardwaj et al. [55] introduced an advanced polymerisation process in which water was added to solid precursor powder obtained by co-grinding raw materials. Amorphous tricalcium phosphate was also added as it was proved that phosphate-based coatings possessed effective corrosion protection and fire protective properties. It could also enhance the inhibition efficiency and improve the passivity of the metal surface. At a coating thickness of about 115 μm, the cold side temperature curve showed that the insulation capacity (time required to reach a temperature of 180°C) was approximately 7 min. This was about three times longer than that of an uncoated steel substrate. It was predicted that the protection efficiency might exceed up to 1 h if the thickness of the coating increased to 20 mm. The direct flame test was also carried out, and the results proved that the increased inorganic phosphate content led to superior fire protective characteristics. The coated sample can withstand a flame for more than 45 min without peeling off. No visual cracks or degradation were observed in the coating after 50 min.

7.3.3 Chemical Resistance

Corrosion in steel structures is a critical issue, especially for infrastructures exposed to marine environments. It had nonnegligible impacts on economic losses, human health, and environmental pollution [7]. Geopolymer coating offers higher strength, longer durability, less permeable mix design, and superior chemical resistance, which can be seen as an excellent alternative to conventional coating materials [4].

Deshmukh et al. [56] demonstrated that geopolymer coating exhibited strong corrosion resistance via electrochemical measurements by dipping the coated mild steel plate in NaCl electrolyte. The coated mild steel plate showed a significantly lower current density (5×10^{-7} A/cm^2) compared to the uncoated mild steel plate (1 A/cm^2). After 72 h of dipping, an insignificant increment in current flow was observed, possibly due to the metal oxide/hydroxide formation.

Bhardwaj [55] discussed the corrosion behaviour of phosphatic geopolymer coating applied on mild steel. The results showed that the coated mild steel exhibited strong corrosion resistance ability even after 20 h of salt spray test, while the uncoated mild steel corroded within 7 h. A non-electrochemical weight loss method was also performed to determine the corrosion rate of coated mild steel. It was demonstrated that the coating containing tricalcium phosphate and sodium metasilicate provided the highest resistance towards the saline condition. Both of these compounds act as anodic inhibitors (also known as passivation inhibitors). The addition of phosphate in the geopolymer matrix makes the coating denser and less porous, so water cannot reach the passivation layer. Thus, the phosphate groups retards corrosion by promoting the growth of protective iron oxide films and by healing the defects in protective films. Meanwhile, the silicon ion forms negatively charged colloidal particles that migrate to anodic areas and form a passive film. These ensure the corrosion protection efficiency of the coating materials.

Tomar et al. [45] evaluated the effect of red mud amount on the corrosion resistance of the geopolymer coating applied on mild metal. The corrosion test was performed in a salt spray chamber in 3.5% NaCl with 95% RH. Results showed that the rust creepage tends to decrease with the addition of red mud (5 to 20 wt%. This could be due to the highly dense and intact structure of geopolymer coating with more red mud amount. It was mentioned previously that the adhesion strength also increased with the increasing amount of red mud, indicating that the involvement of red mud improved the overall property of the coating materials.

To further improve the corrosion resistance, Yang et al. [7] introduced graphene oxide into the geopolymer coating system (Figure 7.4), and the influence of graphene

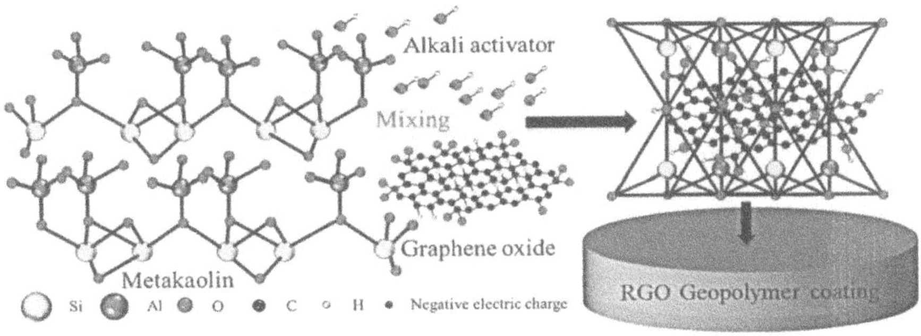

FIGURE 7.4 The schematic illustration of the preparation and coating process for reduced graphene oxide geopolymer coating. (Reprinted with permission from ref. [7]. Copyright 2022 Elsevier and [24].)

oxide contents on the physicochemical and electrochemical properties was investigated. It was indicated that the appearance of graphene oxide showed a noble corrosion potential with very low corrosion density (in the order of 10^{-7}), which shows the effectiveness of graphene oxide geopolymer coating in inhibiting corrosion. In addition, using a graphene oxide content of 0.1% could increase the corrosion resistance of geopolymer-coated steel by more than two orders of magnitude compared to bare carbon steel.

Although the corrosion resistance of geopolymer coating applied on mild steel structures has been investigated widely, it was suggested that the geopolymer coatings were not appropriate for aluminium substrate due to the high alkalinity of the geopolymer, which can corrode the aluminium [47].

7.3.4 Shrinkage

A crack-free coating on substrates is mandatory to sustain its function, whether for fire resistance or anti-corrosive protection. Cracks could occur due to the shrinkage of coatings or the mismatch between the coatings and substrates attained [2, 57]. Therefore, optimising the preparation process or adjusting the recipe for chosen substrates or applications is necessary.

Mao et al. [58] investigated the effect of curing temperature on MK-based geopolymer coating applied on metals by airbrush spraying. Results showed that an adequate microstructure was observed at a moderate curing temperature (80°C). Curing at lower or higher temperatures leads to several or numerous cracks. At low temperatures, several microcracks occurred due to the evaporation of water. At high temperatures, cracks were generally caused by the residual stress between the coating and the substrate. In addition, the difference in the MK-based geopolymer coating on an aluminium substrate was also explored. As the coating was cured at 40°C, plastic deformation was seen, and no cracks were observed within the sample. This could be due to the incomplete polymerisation of the coating. Unlike the coating applied on the mild steel, the coating process failed as the coating was cured at 80°C. As the curing temperature further increased to 150°C, severe cracks were observed from the beginning of the scratch test. As a result, the coating was completely removed from the aluminium substrate by a load of 12 N. The authors suggested that chemical adhesion may have been involved due to the large differences in adhesive strength on mild steel compared with other substrates.

The water content is also a critical factor influencing the shrinkage of the coating. Temuujin et al. [38] studied the effect of water content on the thermal expansion or shrinkage of the geopolymer coating. With less water content, the initial shrinkage of the samples tends to reduce before 200°C, while the expansion of the coatings was high thereafter. Zhang et al. [2] demonstrated that the shrinkage of geopolymer coatings could be improved by adding PP fibre and a MgO expansion agent.

The selection of geopolymer mixture design, curing conditions, and coating technique significantly impact the adhesion strength, thermal and chemical resistance, and microstructure. This has been explored extensively. However, the bonding behaviour (either physical or chemical or both) needs to be further investigated to fully understand the adhesion mechanism between the coating and the substrate. Besides, the majority of the investigations are on a lab scale, where an on-site experiment of the geopolymer coating is necessary to promote a large-scale application.

7.4 THE COMPARISON BETWEEN THE CONVENTIONAL COATING AND GEOPOLYMER COATING

As mentioned previously, geopolymer exhibits excellent thermal stability, relatively high strength, and fire and chemical resistance compared with traditional cement. The study indicated that the geopolymeric materials showed only slight cracks after exceeding 800°C, while the cracks are intensive for the samples made of conventional concrete under fire resistance test. In addition, no spattering was observed for geopolymers at very high temperatures, indicating excellent thermal stability [1, 38, 46].

Phoo-ngernkham et al. [59] discussed the effect of raw materials composition (FA, GGBFS, NaOH, and Na_2SiO_3) on the adhesion strength. Results indicated that the mixture containing FA and GGBFS activated via NaOH and Na_2SiO_3 combined solution possessed the maximum shear bond strength (Figure 7.5). It is also worth mentioning that the shear bond strength of the geopolymer coating was 19–47% higher than the two common organic coatings (epoxy A and B). Zhang et al. [2] compared the bonding strength between geopolymer coating and cement-based materials at various drying conditions. Results showed that the 1-day bonding strength was equal to or higher than that of cement-based materials at any chosen curing condition (dry-wet cycle curing, air-curing, and seawater-curing).

Lahoti et al. [60] compared the fire performance of Portland cement concrete against geopolymer concrete. It was stated that the thermal conductivity of geopolymer (0.2–0.4 W/m.K) is generally much lower than cement concrete (1–4 W/m.K), although both materials are incombustible. Geopolymer concrete also exhibits better spalling resistance compared to OPC concrete due to the interconnected pore structure of geopolymers. More importantly, geopolymer concrete provides excellent fire resistance properties up to 1200°C, while a significant strength loss was observed for OPC concrete after 600°C. At high temperatures, geopolymers maintained microstructural stability as the chemical bonds in geopolymers do not break

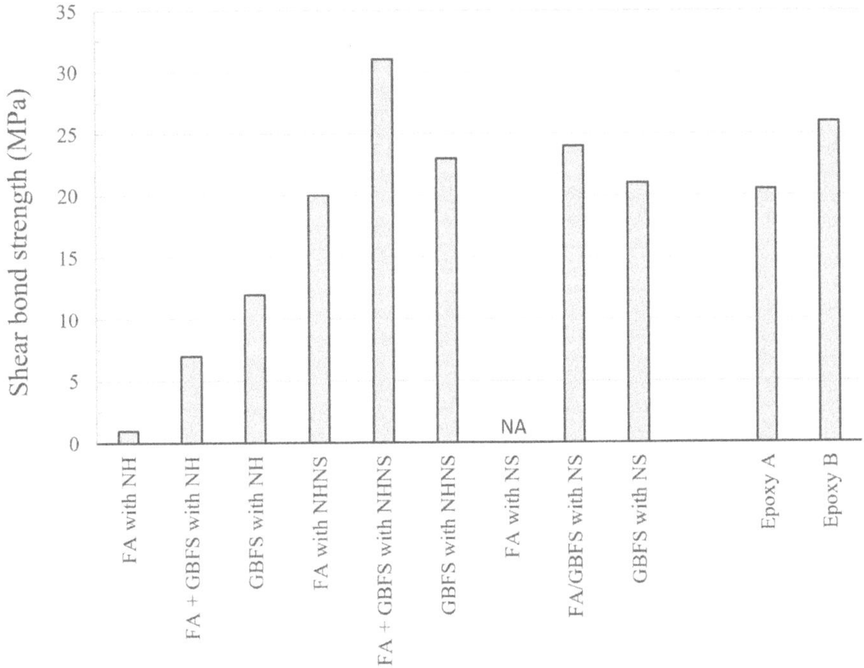

FIGURE 7.5 Shear bond strength between geopolymer paste, epoxy, and substrate with interface line at 45° to the vertical. (Reprinted with permission from ref. [59]. Copyright 2015 Elsevier and [24].)

down, although a transformation from the amorphous phase to the crystalline phase is generally observed at high temperatures (close to 1000°C). In contrast, cement paste and aggregates lose their strength above 600°C, indicating that the concrete is considered structurally failed.

Wang et al. [61] evaluated the environmental impacts of geopolymer coatings against Portland cement coating via life cycle assessment (LCA). LCA can identify the main contributors to environmental impacts and assist in decision-making by pointing out potential aspects for future improvement. In addition, the effect of fabrication parameters on the overall environmental performance was also addressed. It was concluded that the geopolymer coating yielded substantially lower environmental impacts (50–80% lower in most impact categories) than OPC coating. For cement-based coating, sand and cement are the two major contributors to the environmental impacts due to the limestone's calcination and the high temperature needed to heat the raw materials. As a result, the estimated CO_2 emissions during cement concrete production could increase up to 45% more than geopolymer concrete [30]. For geopolymer coating, waterglass contributes considerably to the overall environmental impacts as waterglass production requires heating at high temperatures and emits heavy metal ions into the water environment (Figure 7.6). In addition, an increase in the W/S ratio decreased the environmental impacts, and slag-based geopolymer coating achieved lower environmental impacts than FA-based and MK-based

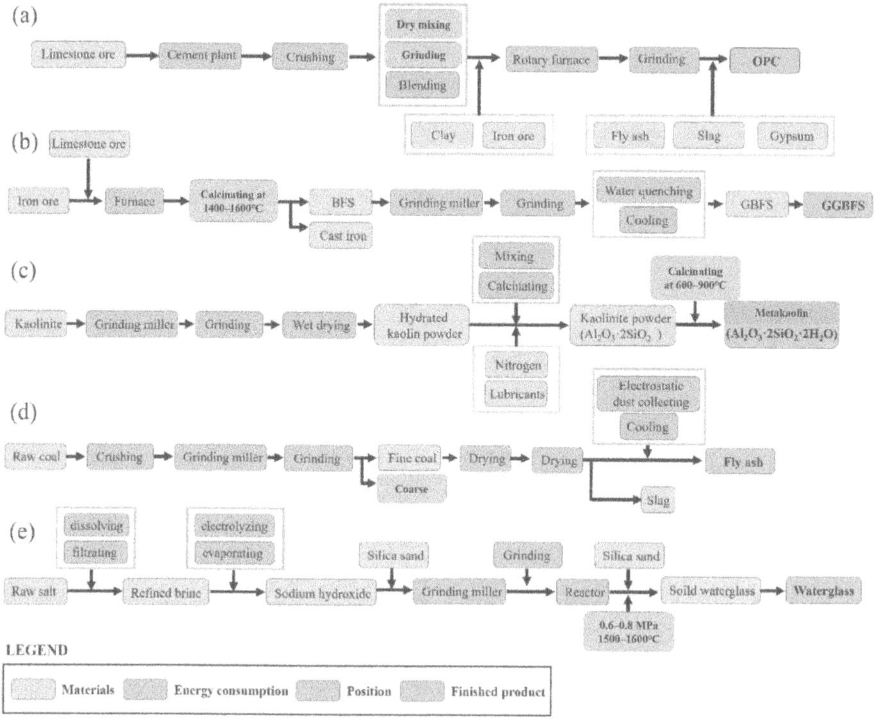

FIGURE 7.6 Life-cycle boundary of (a) ordinary Portland cement, (b) slug, (c) MK, (d) FA, and (e) waterglass [24, 61].

varieties. These findings provide useful guidance for balancing the performance and environmental impacts of geopolymer coating [61].

Geopolymer coating exhibits superior mechanical and thermal performance compared with conventional cement-based coating. The life cycle analysis between these two coatings indicates that the geopolymer coating has a substantially lower environmental impact than cement-based coating. Apart from these investigations, an economic analysis of geopolymer coating could be carried out and compared with current commercial protective coatings. It was suggested that the benefit-to-cost ratio is a more precise index to utilise in the economic analysis [4].

7.5 CONCLUSIONS

This chapter reviewed the current use of conventional passive coating and assessed the potential use of novel geopolymer coating on the metal substrate, practically, the steel structure. The influencing factors, including precursor materials, the alkaline activator, curing processes, and water contents, on the adhesion, thermal, and chemical resistance of the geopolymer coating were well explored. The performance comparison between geopolymer coating and conventional cement-based coating indicated that geopolymer coating offered superior mechanical and thermal performance with substantially lower environmental impact compared with cement-based coating.

Meanwhile, further study is still required. Although the bonding strength has been widely investigated, the mechanism of adhesion between the coating and the substrate is still unclear. Therefore, the bonding behaviour (physical, chemical, or both) needs further investigation. In addition, the required thickness of the fireproof coating is much higher than that of intumescent coating, which increases the structural self-weight. Thus, the trade-off between the weight and the thermal-mechanical performance needs to be identified.

Moreover, the preparation and production of geopolymer mainly rely on aluminosilicate precursors and activators. Due to the absence of technical standards and the diversity of constituent materials, the properties of the geopolymer are less controllable compared with cement-based materials. Therefore, further standardisation of the raw material and the fabrication process is needed to achieve custom-tailored products for particular performances and applications.

The life cycle analysis indicates that the geopolymer coating has a substantially lower environmental impact compared with cement-based coating. Apart from this, an economic analysis of geopolymer coating could also be carried out and compared with current commercial protective coatings. Besides, on-site experiments on geopolymer coating are necessary to verify all the factors investigated. Nevertheless, geopolymer coatings show great potential for fire protection on steel structures with excellent physical properties, lower environmental impact, and cost savings.

ACKNOWLEDGEMENTS

The financial support of Tsinghua-Foshan Advanced Manufacturing Research Institute is acknowledged. The authors are grateful for the support of colleagues at the Future Materials & Design Research Center, The Future Lab, Tsinghua University. The assistance of the technical staff at the School of Materials Science and Engineering Analysis Center, Tsinghua University is also acknowledged.

REFERENCES

1. K. Kaczmarski, K. Pławecka, B. Kozub, P. Bazan, and M. Łach, "Preliminary investigation of geopolymer foams as coating materials," *Applied Sciences*, vol. 12, no. 21, 2022.
2. Z. Zhang, X. Yao, and H. Zhu, "Potential application of geopolymers as protection coatings for marine concrete. I. Basic properties," *Applied Clay Science*, vol. 49, no. 1–2, pp. 1–6, 2010.
3. L. Andrea and M. Cristian, "Intumescent coatings used for the fire-safe design of steel structure: A review," *Journal of Constructional Steel Research*, vol. 162, 2019.
4. L. S. Soong, M. Y. J. Liu, S. P. Yap, K. H. Mo, M. Z. Jumaat, and Y. Goh, "The potential of geopolymer in development of green coating materials: A review," *Arabian Journal for Science and Engineering*, vol. 47, no. 10, pp. 12289–12299, 2022.
5. R. G. Puriand A. S. Khanna, "Intumescent coatings: A review on recent progress," *Journal of Coatings Technology and Research*, vol. 14, no. 1, pp. 1–20, 2017.
6. D. de Silva, I. Nuzzo, E. Nigro, and A. Occhiuzzi, "Intumescent coatings for fire resistance of steel structures: Current approaches for qualification and design," *Coatings*, vol. 12, no. 5, 2022.
7. N. Yang, C. S. Das, X. Xue, W. Li, and J.-G. Dai, "Geopolymer coating modified with reduced graphene oxide for improving steel corrosion resistance," *Construction and Building Materials*, vol. 342, 2022.

8. Q. Tian, S. Wang, Y. Sui, and Z. Lv, "Alkali-activated materials as coatings deposited on various substrates: A review," *International Journal of Adhesion and Adhesives*, vol. 110, 2021.
9. Q. Wei, Y. Liu, and H. Le, "Mechanical and thermal properties of phosphoric acid activated geopolymer materials reinforced with mullite fibers," *Materials (Basel)*, vol. 15, no. 12, 2022.
10. Y. Fang *et al.*, "Property evolution of geopolymer composites with SiC whiskers loaded with BN coating at elevated temperatures," *Construction and Building Materials*, vol. 309, 2021.
11. L. K. Turner and F. G. Collins, "Carbon dioxide equivalent (CO2-e) emissions: A comparison between geopolymer and OPC cement concrete," *Construction and Building Materials*, vol. 43, pp. 125–130, 2013.
12. J. Xu, Y. Zhou, and X. Tang, "Study on the solidication of heavy metals by fly ash based geopolymers" (in Chinese), *Journal of Building Materials*, vol. 9, no. 3, 2006.
13. F. F. Zainal, M. F. Fazill, K. Hussin, A. Rahmat, M. M. A. B. Abdullah, and W. Wazien, "Effect of geopolymer coating on mild steel," *Solid State Phenomena*, vol. 273, pp. 175–180, 2018.
14. X. Ma, J. Pan, J. Cai, Z. Zhang, and J. Han, "A review on cement-based materials used in steel structures as fireproof coating," *Construction and Building Materials*, vol. 315, 2022.
15. B. Yin *et al.*, "Research and application progress of nano-modified coating in improving the durability of cement-based materials," *Progress in Organic Coatings*, vol. 161, 2021.
16. A. Noumowe, "Mechanical properties and microstructure of high strength concrete containing polypropylene fibres exposed to temperatures up to 200°C," *Cement and Concrete Research*, vol. 35, no. 11, pp. 2192–2198, 2005.
17. L. Sarvaranta and E. Mikkola, "Fibre mortar composites in fire conditions," *Fire and Materials*, vol. 18, no. 1, pp. 45–50, 1994.
18. K. Pierre, C. Grégoire, and G. Christophe, "High-temperature behaviour of HPC with polypropylene fibres from spalling to microstructure," *Cement and Concrete Research*, vol. 31, pp. 1487–1499, 2001.
19. J. H. Lee, Y. S. Sohn, and S. H. Lee, "Fire resistance of hybrid fibre-reinforced, ultra-high-strength concrete columns with compressive strength from 120 to 200 MPa," *Magazine of Concrete Research*, vol. 64, no. 6, pp. 539–550, 2012.
20. T. Mariappan, "Recent developments of intumescent fire protection coatings for structural steel: A review," *Journal of Fire Sciences*, vol. 34, no. 2, pp. 120–163, 2016.
21. L. Wang, Y. Dong, C. Zhang, and D. Zhang, "Experimental study of heat transfer in intumescent coatings exposed to non-standard furnace curves," *Fire Technology*, vol. 51, no. 3, pp. 627–643, 2015.
22. S. M. Anees and A. Dasari, "A review on the environmental durability of intumescent coatings for steels," *Journal of Materials Science*, vol. 53, no. 1, pp. 124–145, 2017.
23. J. Hao and W. K. Chow, "A brief review of intumescent fire retardant coatings," *Architectural Science Review*, vol. 46, no. 1, pp. 89–95, 2003.
24. K. Wang and H. Le, "The development of cement-based, intumescent and geopolymer fire-retardation coatings for metal structures: A review," *Coatings*, vol. 13, no. 3, 2023.
25. C. Réti, M. Casetta, S. Duquesne, S. Bourbigot, and R. Delobel, "Flammability properties of intumescent PLA including starch and lignin," *Polymers for Advanced Technologies*, vol. 19, no. 6, pp. 628–635, 2008.
26. M. C. Yew, N. H. Ramli Sulong, M. K. Yew, M. A. Amalina, and M. R. Johan, "Influences of flame-retardant fillers on fire protection and mechanical properties of intumescent coatings," *Progress in Organic Coatings*, vol. 78, pp. 59–66, 2015.
27. G. Samuel, "Intumescent fire-retardant composition for high temperature and long duration protection," US Patent US5723515 A, 1995.

28. J. H. Koo,, P. S. Ng, and C. F.-B. Cheung, "Effect of high temperature additives in fire resistant materials," *Fire Sciences*, vol. 15, no. 6, pp. 427–504, 1997.
29. B. Du and Z. Fang, "Effects of carbon nanotubes on the thermal stability and flame retardancy of intumescent flame-retarded polypropylene," *Polymer Degradation and Stability*, vol. 96, no. 10, pp. 1725–1731, 2011.
30. K. Wang, "Novel development of eco-friendly porous thermal insulation materials and the application," PhD thesis, Mechanical and Aerospace Engineering, University of Strathclyde, 2022.
31. C. Jiang, A. Wang, X. Bao, T. Ni, and J. Ling, "A review on geopolymer in potential coating application: Materials, preparation and basic properties," *Journal of Building Engineering*, vol. 32, 2020.
32. J. Temuujin, A. Minjigmaa, W. Rickard, M. Lee, I. Williams, and A. van Riessen, "Preparation of metakaolin based geopolymer coatings on metal substrates as thermal barriers," *Applied Clay Science*, vol. 46, no. 3, pp. 265–270, 2009.
33. Z. Zhang, X. Yao, and H. Wang, "Potential application of geopolymers as protection coatings for marine concrete III. Field experiment," *Applied Clay Science*, vol. 67–68, pp. 57–60, 2012.
34. B. Lu *et al.*, "Study of MgO-activated slag as a cementless material for sustainable spray-based 3D printing," *Journal of Cleaner Production*, vol. 258, 2020.
35. J. Liyana, H. Kamarudin, A. M. Mustafa Al Bakri, M. Binhussain, C. M. Ruzaidi, and A. M. Izzat, "," *Australian Journal of Basic and Applied Sciences*, vol. 7, no. 5, pp. 182–186, 2013.
36. G. Fahim Huseien, J. Mirza, M. Ismail, S. K. Ghoshal, and A. Abdulameer Hussein, "Geopolymer mortars as sustainable repair material: A comprehensive review," *Renewable and Sustainable Energy Reviews*, vol. 80, pp. 54–74, 2017.
37. P. Xiaoying, S. Zhenguo, S. Caijun, L. Tung-Chai, and L. Ning, "A review on concrete surface treatment Part I: Types and mechanisms," *Construction and Building Materials*, no. 132, pp. 578–590, 2017.
38. J. Temuujin, A. Minjigmaa, W. Rickard, M. Lee, I. Williams, and A. van Riessen, "Fly ash based geopolymer thin coatings on metal substrates and its thermal evaluation," *Journal of Hazardous Materials*, vol. 180, no. 1–3, pp. 748–52, 2010.
39. M. Irfan Khan, K. Azizli, S. Sufian, and Z. Man, "Sodium silicate-free geopolymers as coating materials: Effects of Na/Al and water/solid ratios on adhesion strength," *Ceramics International*, vol. 41, no. 2, pp. 2794–2805, 2015.
40. X. Rong, Z. Wang, X. Xing, and L. Zhao, "Review on the adhesion of geopolymer coatings," *ACS Omega*, vol. 6, no. 8, pp. 5108–5112, 2021.
41. X. Lv, K. Wang, Y. He, and X. Cui, "A green drying powder inorganic coating based on geopolymer technology," *Construction and Building Materials*, vol. 214, pp. 441–448, 2019.
42. P. Duxson, A. Fernández-Jiménez, J. L. Provis, G. C. Lukey, A. Palomo, and J. S. J. van Deventer, "Geopolymer technology: The current state of the art," *Journal of Materials Science*, vol. 42, no. 9, pp. 2917–2933, 2006.
43. M.S. Siti Salwa, A. M. Mustafa Al Bakri, H. Kamarudin, C. M. Ruzaidi, M. Binhussain, and S. Z. Sharifah Zaliha, "Review on current geopolymer as a coating material," *Australian Journal of Basic and Applied Sciences*, vol. 5, no. 7, pp. 246–257, 2013.
44. G. C. Tatlisu, C. Aciksari, S. Celebi, and S. Turan, "Developing a hollow glass microsphere/geopolymer thermal insulation composite for hot metal surface coating," *Ceramics International*, vol. 48, no. 9, pp. 11924–11939, 2022.
45. A. Singh Tomar, R. Gupta, A. Singh, S. Thankaraj Salammal, M. Akram Khan, and D. Mishra, "Evaluation of corrosion protective properties of fly ash-red mud based geopolymer coating material for mild steel," *Materials Today: Proceedings*, vol. 68, pp. 181–186, 2022.

46. S. L. Yong, D. W. Feng, G. C. Lukey, and J. S. J. van Deventer, "Chemical characterisation of the steel–geopolymeric gel interface," *Colloids and Surfaces A: Physicochemical and Engineering Aspects*, vol. 302, no. 1–3, pp. 411–423, 2007.
47. B. Jonathan, G. Matthew, and K. Waltraud, *Use of geopolymeric cements as a refractory adhesive for metal and ceramic joins* (Advances in Ceramic Coatings and Ceramic-Metal Systems: Ceramic Engineering and Science Proceedings). The American Ceramic Society, 2005.
48. B. A. Latella, D. S. Perera, T. R. Escott, and D. J. Cassidy, "Adhesion of glass to steel using a geopolymer," *Journal of Materials Science*, vol. 41, no. 4, pp. 1261–1264, 2006.
49. S. De Barros, J. R. De Souza, K. C. Gomes, E. M. Sampaio, N. P. Barbosa, and S. M. Torres, "Adhesion of geopolymer bonded joints considering surface treatments," *The Journal of Adhesion*, vol. 88, no. 4–6, pp. 364–375, 2012.
50. Z. Wang, X. Rong, L. Zhao, X. Xing, and H. Ma, "Effects of substrate surface characteristics on the adhesion properties of geopolymer coatings," *ACS Omega*, vol. 7, no. 14, pp. 11988–11994, 2022.
51. J. Temuujin, W. Rickard, M. Lee, and A. van Riessen, "Preparation and thermal properties of fire resistant metakaolin-based geopolymer-type coatings," *Journal of Non-Crystalline Solids*, vol. 357, no. 5, pp. 1399–1404, 2011.
52. T. Bakharev, "Thermal behaviour of geopolymers prepared using class F fly ash and elevated temperature curing," *Cement and Concrete Research*, vol. 36, no. 6, pp. 1134–1147, 2006.
53. A. I. Nicoară, A. I. Bădănoiu, G. Voicu, C. Dinu, and A. Ionescu, "Intumescent coatings based on alkali-activated borosilicate inorganic polymers," *Journal of Coatings Technology and Research*, vol. 17, no. 3, pp. 681–692, 2019.
54. J. Sarazin *et al.*, "Flame resistance of geopolymer foam coatings for the fire protection of steel," *Composites Part B: Engineering*, vol. 222, 2021.
55. P. Bhardwaj, R. Gupta, D. Mishra, S. K. Sanghi, S. Verma, and S. S. Amritphale, "Corrosion and fire protective behavior of advanced phosphatic geopolymeric coating on mild steel substrate," *Silicon*, vol. 12, no. 3, pp. 487–500, 2019.
56. K. Deshmukh, R. Parsai, A. Anshul, A. Singh, P. Bharadwaj, R. Gupta, and D. Mishra, "Studies on fly ash based geopolymeric material for coating on mild steel by paint brush technique," *International Journal of Adhesion and Adhesives*, no. 75, pp. 139–144, 2017.
57. Z. Zhang, X. Yao, and H. Zhu, "Potential application of geopolymers as protection coatings for marine concrete. II. Microstructure and anticorrosion mechanism," *Applied Clay Science*, vol. 49, no. 1–2, pp. 7–12, 2010.
58. Y. Mao, L. Biasetto, and P. Colombo, "Metakaolin-based geopolymer coatings on metals by airbrush spray deposition," *Journal of Coatings Technology and Research*, vol. 17, no. 4, pp. 991–1002, 2020.
59. T. Phoo-ngernkham, A. Maegawa, N. Mishima, S. Hatanaka, and P. Chindaprasirt, "Effects of sodium hydroxide and sodium silicate solutions on compressive and shear bond strengths of FA–GBFS geopolymer," *Construction and Building Materials*, vol. 91, pp. 1–8, 2015.
60. M. Lahoti, K. H. Tan, and E.-H. Yang, "A critical review of geopolymer properties for structural fire-resistance applications," *Construction and Building Materials*, vol. 221, pp. 514–526, 2019.
61. A. Wang, Y. Fang, Y. Zhou, C. Wang, B. Dong, and C. Chen, "Green protective geopolymer coatings: Interface characterization, modification and life-cycle analysis," *Materials (Basel)*, vol. 15, no. 11, 2022.

8 Geopolymers in the Marine Environment

Kinga Korniejenko, Mikuła Janusz, and Kozub Barbara

8.1 INTRODUCTION

The aquatic environment places many demands on materials to be used for applications in seas and oceans [1, 2]. These materials have to be characterized by suitable physical and mechanical properties, durability, long-term resistance against corrosion, and other features dedicated to particular applications. Moreover, they have to be safe for marine fauna and flora [3, 4]. One of the basic challenges is the corrosion and deterioration of materials (Figure 8.1).

FIGURE 8.1 Deterioration of coastal infrastructure in a different type of seas and oceans: (a) harbor in Athens, Greece; (b) seaside promenade in Nha Trang, Vietnam; (c) coastal infrastructure in Tel Aviv, Israel; (d) pier in Gdansk, Poland.

DOI: 10.1201/9781003535683-8

Materials for underwater applications should be durable and resistant to marine environments [4]. The coastal regions are focusing on the challenges of maintaining the sustainability of marine resources, limiting damage to the ecosystem, and protecting against floods and coastal erosion. The cost of corrosion of different water infrastructure, including offshore facilities, is estimated to be approximately 4% of GDP [5]. Basic problems are corrosion and deterioration of the materials. Despite the resistance to corrosion, materials can be affected by environmental factors [4]. Furthermore, the presence of soluble salts from greenhouse acid gases and other water pollution, such as sulphur oxide, nitric oxide, and carbon oxide, can accelerate the corrosion process [6]. The deterioration of a material is not only an aesthetic problem; the pollutants can weaken structural barriers, accelerating corrosion. In marine environment, the main cause of corrosion in concrete structures is the intrusion of chlorides [7, 8], sodium chlorides, and sodium sulphates [9]. Materials, in spite of their resistance to corrosion, can be damaged due to the influence of environmental stressors [9, 10]. On the surface, it is caused by exposure to weather agents such as wind, temperature, rain, relative humidity, and condensation. These agents induce physical and chemical reactions, which destroy the material and cause chemical corrosion or erosion by oxidation and dehydration reactions, dissolution of carbonates, and dissolution of some minerals containing silicates. In the bulk of concrete structures, the crystallization of salts in the pores of materials can cause dimensional movements that result in cracking [11, 12], in addition to the corrosion due to chloride ingress, sodium chlorides, and sodium sulphates [13]. Furthermore, in this environment, a distinctive problem is bioerosion. It is mainly a chemical process, but it may be assisted by mechanical means [14] – by bio abrasion and biocorrosion. It is a significant problem, especially in underwater applications. A wide range of materials can suffer physical, chemical, and aesthetic damage from insects, algae, lichens, fungi, and bacteria [14, 15]. The surfaces of such materials are used as substrates by a wide variety of microorganisms, chemoorganotrophs, chemolithotrophs and phototrophs, actinomycetes, fungi, and lichens [16]. The intensity of biological attack and biodeterioration processes is strongly influenced by water availability. Biodeterioration is determined by: material parameters (porosity, permeability), their conservation status (low ductility, erosion, microcracks), environmental conditions, structure morphology, and interactions between the support and the environment over time [16, 17]. They must be resistant not only to salt water but also to water pollution (in inland and marine waters).

According to current research, geopolymers can be highly desirable for marine applications due to their physio-chemical properties, particularly long-term stability and resistance to different environments [2, 10]. The preliminary research shows the successful application of geopolymers for marine applications such as repair works [11–13]. Current building materials are environmentally damaging: concrete, for example, is considered toxic to many marine organisms and 70% of all coastal infrastructure. This toxicity is due to high surface alkalinity (pH 13 compared to pH 8 of seawater) and leaching of heavy metals. Therefore, there is a need to find new materials and technologies for application in marine environments.

The main aim of this chapter is to analyse the current progress in developing geopolymers for marine applications as a potentially interesting material with a limited environmental burden. The study is based on the systematic review that was performed using Scopus as the main search tool and the supportive tools in the following databases: ScienceDirect, IEEE Xplore Digital Library, ACS Publications, Wiley Online Library, and Google Scholar. It was conducted between August and September 2024. The keywords used were "geopolymer" and "marine" combined. The research results in the Scopus database showed 172 documents. They were checked, and the most relevant publications were selected for this review. The analysis of the literature showed that the topic was very new. The first publication was in 2014, and after 2019, the growth of interest in this area was noted. In conclusion, from the first look at this data, it can be noted that the research area is innovative, and it has been intensively developed in the last few years. This chapter addresses researchers and practitioners who are demonstrating the state of the art in the field of geopolymer applications in marine environments. It can be helpful for this group to define the attractive research fields in this area. This chapter also defines the most important trends for the further development of geopolymer materials for marine applications.

8.2 MATERIAL COMPONENTS

8.2.1 Marine-Based Ingredients as a Replacement for Traditional Components of Geopolymers

The first field of investigation for geopolymers in marine environments is proper raw materials that ensure the necessary properties and will be cost-effective as well as safe for the natural environment. Currently, the most popular raw materials for geopolymer manufacturing are metakaolin and industrial by-products such as fly ash and furnace slags [18]. These materials quite often require some modifications to improve their properties, such as corrosion resistance and durability. The alternative approach is to use the materials that come from marine sources. This increased the chance that there will be more durability in this environment. This kind of research was made by Pradhan *et al.* [19], who investigated a one-part geopolymer manufactured from the quartz-rich marine clay available in Singapore [19]. The results show that marine clay is not ideal for a precursor material in geopolymerization because of the low amount of kaolin in the raw material (~29%) and the necessity to activate in high temperatures (calcination at 850°C) or with using the milling process. However, it can be applied as an alternative raw material for the marine environment [19].

The other tested raw materials were sediments [20, 21]. The main aim of the provided research was to increase the economy of dredging practices. The dredged sediments are a potential source of material, but because of the unfavorable properties, especially high setting times and shrinkage, they are not possible to use as only one material [21]. This type of raw material requires to be mixed with other materials, such as concrete or alkali reagents, to obtain useful material for civil engineering applications [20, 21]. The trials in this area were made for the co-valorization of marine sediments using metakaolin and ground-granulated blast furnace slag (GGBFS) to obtain sediment-based geopolymer mortars [21] and with coal fly ash and ground-granulated blast furnace slag to obtain low-cost lightweight artificial aggregates [20].

The other important element for the design of geopolymer material is aggregate. Geopolymer concrete with fine aggregate is commonly used for marine infrastructure. To create an effective solution to this environment, it is possible to use recycled materials, such as industrial by-products and construction and demolition waste. The investigation provided by Irum and F. Shabbir [22] shows that this kind of aggregate could have a positive influence on the reduction of environmental burden, but it negatively influences material properties, such as mechanical strength and durability [22]. Because of that the application of this kind of aggregates is limited.

The other possibility is the incorporation of coral aggregate in geopolymer concrete. It allows for cost and time savings associated with the transportation of raw materials, especially if the place of construction is located in the offshore region [23]. Additionally, this approach allows for reducing carbon emissions. The research also shows that the addition of this kind of sand allows the material to resist the sulphate ions erosion in the marine environment [23]. Similar results were achieved also by Anbarasan *et al.* [24]. This team confirms that dredged marine sand can be used as a fine aggregate to make geopolymer concrete. This kind of geopolymer in terms of corrosion resistance, carbonation, and alkalinity has comparable properties as geopolymers manufactured with the usage of river sand [24]. The sea sand was also the subject of an investigation by Xu *et al.* [25]. They also confirmed that sea sand had a positive effect on geopolymer-based material in the area of corrosion resistance [25].

Moreover, the usage of marine sand as an ingredient of geopolymer materials can positively influence their properties, such as compressive and flexural strength, giving the material superior structural durability and environmental friendliness [26]. The other type of the origin material that can be applied in coastal areas is coral waste. Yang *et al.* [27] applied coral waste as the internal curing material to prepare high-performance marine geopolymer concrete. Moreover, they combine this material with other marine-based materials such as seawater and sea sand [27]. This research shows that coral coarse aggregate could be a valuable addition to the amount of between 40% and 50% [27]. The potential disadvantages of using coarse coral aggregate can also be connected to the high porosity and brittleness of this type of aggregate, which affect the mechanical characteristics and durability of the material [28]. However, in practice, the research results prove that dense microstructure and stabilized hydration products of geopolymers positively influence corrosion resistance in seawater compared to traditional cement-based materials [28].

The materials, such as sea sand and coral wastes, were also tested together as environmentally sustainable marine-engineered geopolymer composites [29]. As a result, the flowability and drying shrinkage decrease. The behavior of the compressive strength was changeable in time, with an initial rise but finally a decrease. The marine-based components have positive influence on the tensile ductility. These phenomena were caused by changes in microscopic structure, especially hydration products and pore structure [29]. Another marine-based material that was tested in the application of geopolymer materials is marine shell waste [30]. This waste was used as a replacement for a part of metakaolin as a raw material for geopolymer production. The results showed that adding shell waste reduced the compressive strength of geopolymer materials and slightly increased the elasticity modulus [30]. The reduction of mechanical properties is compensated by environmental and economic benefits. The provided analysis showed that replacing metakaolin with an average amount of 10% seashell waste reduces energy usage by 7% and the price by 2% to produce 1 ton of geopolymer [30]. Moreover, this kind of activity prevents the depletion of non-renewable sources of kaolin and solves the problems related to seashell waste landfilling [30].

Another important resource for building material manufacturing is water. In the coastal area, huge amounts of salt water are usually available, but freshwater is not always accessible. Because of that, the important topic of the research is the possibility of replacing freshwater with salt water replacement. The existing works show a potential to apply salt water to enhance binder systems as well as environmental benefits, including reduction of carbon dioxide emission [31]. The use of seawater was a main research topic suggested by Li *et al.* [32]. They confirm that seawater does not significantly influence the geopolymer microstructure and mechanical properties, and the influence of the ions on geopolymer formation is limited [32]. Also, the ions in salt water influence the geopolymer composition, resulting in the formation of some salts, which was the main reason for reducing its compressive strength [32]. Additionally, the chlorine was distributed in the geopolymer matrix and bounded inside, but SO_4^{2-} underwent precipitation. Overall, the results showed that salt water can be applied for geopolymers used as building materials in the marine environment [32].

Also, some research works combine different marine-based materials, such as salt water and marine sand. This kind of approach was applied by Yang *et al.* [33]. In this case, an addition of salt water caused the formation of magnesium aluminosilicate hydrates, magnesium silicate hydrates, and silica gels filled the pores, which positively affected the mechanical properties [33]. One disadvantage of using salt water is the greater shrinkage of the material during the manufacturing process [33]. Other research on the usage of seawater and sea sand shows that using these components makes it possible to obtain high-performance building materials. The developed geopolymers achieved compressive strength over 140 MPa and high tensile ductility [34]. This research also confirmed that the application of seawater caused a slight reduction of compressive strength. The main reason for this phenomenon is probably the high content of hydrotalcite phases in the material created with the usage of sea seawater [34]. The developed material also showed some advantages compared to material based on freshwater, such as slightly better tensile strength, superior crack resistance, and better sustainability [34]. It is worth noticing that most of the ingredients of geopolymers can be replaced by marine-based components, as shown in Table 8.1.

TABLE 8.1
Possibility of Replacing Traditional Ingredients of Geopolymer by Marine-Based Components

	The Basic Ingredient for Traditional Geopolymer	Marine-Based Component for Replacing	Source
1	Aluminosilicate sources, for example, metakaolin or fly ash	Marine clay, sediments	[19, 20, 21]
2	Fine aggregate, for example, river sand	Marine sand, coral waste	[23, 24, 25]
3	Water	Salt water	[31, 32, 34]
4	Alkaline activator	–	–

Using marine-based resources as the ingredients of materials dedicated to construction industries can significantly increase eco-friendly building and sustainable development in the island and coastal areas. The application of ingredients such as seawater, coral sand or sea sand, or coral coarse aggregate for the preparation of geopolymer concrete could reduce construction time and costs for offshore projects [28]. Also, other components have been tested as potential materials for geopolymerization. Some studies were conducted using *Posidonia oceanica* sea balls (egagropili), which are marine wastes that accumulate along Mediterranean beaches [35]. The first step was to recover halloysite nanotubes (HNTs) and cellulose from this waste and use it to develop hybrid films. Next, the films were used for geopolymers manufacturing by adding NaOH solution for the alkaline activation of nano clay [35]. The geopolymerization process enhances the rigidity of the material. The obtained films can be promising solutions for packaging applications and also as films in building applications [35].

8.2.2 Reinforcement

The important element of most building construction is reinforcement. This element is a key for the most demanding applications, such as bridges. Although steel reinforcement has a longer life in geopolymer concrete than traditional concrete, it is still limited [36]. Because of that, new reinforcement methods for marine construction are required [37, 38]. Currently, basalt fibres are one of the most promising fibres for construction reinforcement in the marine environment. The long-term durability of 1% basalt fibre-reinforced geopolymer concrete was investigated by Zhang *et al.* [39]. The material was subjected to dry-wet cycles and immersion treatment in marine conditions. The results show that the addition of basalt fibres effectively enhances material stability, including mechanical properties and porosity [39]. Other possible benefits connected with basalt fibre reinforcement are improvement in the ultimate carrying capacity of beams and columns and increasing flexural and compressive capacity in comparison to conventional steel-reinforced counterparts [26]. What may also seem interesting in terms of practical applications is the fact that using seawater increases fibre/matrix bond strength [34]. These results were also confirmed through the investigation by Rahman and Al-Ameri [40]. The research was based on geopolymers reinforced by basalt bars, and the experiment was supposed to replicate tidal cycles. The experimental results revealed that the reinforced material shows better resistance in the marine environment, including lower micro- and macro-mechanical degradation and a higher residual ultimate load [40]. Investigation with reinforcement made from basalt bars was also provided by Zhao *et al.* [41]. This research shows some danger connected with the usage of this material in highly alkaline solutions that are harmful in the long term for basalt bars [41].

The other type of research used in geopolymer matrix is glass fibre, which also shows good durability and mechanical properties in marine environments and can be applied in marine engineering [42]. It is also worth mentioning that not all research has positive results. Lu *et al.* [37] investigated different types of fibre reinforcement in a geopolymer matrix created with the usage of seawater and sea sand, as well as fly ash and ground granulated blast furnace slag [37]. This study explores

TABLE 8.2
The Most Popular Reinforcement Types Used in Geopolymers Are Dedicated to Marine Infrastructure Applications

	Kind of Fibres	Type of Fibres	Source
1	Basalt	Bars	[37, 40]
2	Carbon	Bars	[37]
3	Glass	Bars	[37]
4	Basalt	Short fiber	[39]

the reinforcement in the form of glass, basalt, and carbon bars and the influence of replacement river sand by sea sand on these compositions. In this case, the long-term performances of the composites based on both types of sand have similar properties and were characterized by reduced strength and enhanced brittleness in seawater [37]. The bonding strength between all types of bars and matrix decreased due to the decreased concrete strength under seawater immersion, and fibre fractured under the effects of sustained load [37]. In marine applications, there is a visible trend to find new types of reinforcement for the replacement of traditional steel bars to avoid corrosion problems (Table 8.2).

The new types of applications are based mainly on organic polymers and mineral fibres that are characterized by high resistance to marine corrosive environments. The replacement of traditional steel bars with other fibres allows for the enhancement of the corrosion resistance and durability of the construction in salt water.

8.2.3 Other Additives Applied to Reinforce Selected Properties in the Marine Environment

The other important area of investigation is additives and admixtures for geopolymers to enhance selected characteristics in marine environments. Among the different materials that have been tested, interesting results have been obtained using nano fly ash particles [43]. This kind of filler, in addition of 1%, acts as a precursor of the geopolymerization reaction, resulting in a denser geopolymer medium and higher mechanical properties as well as higher corrosion resistance, especially a negligible level of chloride ion penetration [43]. Another approach to waterproof improvement is the different additives based on polymer compounds. One of the investigated additives was polydimethylsiloxane (PDMS) [44]. The results showed that the addition of PDMS efficiently support the waterproof properties, such as increasing the contact angle of geopolymer composites, giving the material hydrophobic surface wettability, and decreasing the water absorption. Moreover, the PDMS improves the compressive properties of geopolymer [44]. Another approach to increase the material's durability is using protective layers or coatings. Geopolymeric materials have also been investigated in this area. The geopolymer is usually resistant to biological pollution, and its high alkalinity gives an anti-fouling effect. Additives such as triethoxycaprylylsilane (TTOS) can be used to enhance the superhydrophobic effect [45]. Based on this surface modification, the geopolymer materials obtained have micro/nanostructures with long-lasting controlled release of Cu^{2+}. Such modifications extend the anti-biofouling properties of geopolymer [45]. Another approach was presented by Phiangphimai *et al.* [46]. They designed a coating made from alkali-activated/cement powder paste (geopolymer powder, Portland cement, and silica fume) for the improved durability of marine structures. The results showed good resistance of such coating against reactive environments such as 5% sulphuric acid (H_2SO_4) solution, 5% magnesium sulphate ($MgSO_4$) solution, tap water (H_2O), and chloride migration coefficient. The proposed coating seems to be a good solution for preventing corrosion in marine environments [46].

8.3 KEY PROPERTIES IN THE MARINE ENVIRONMENT

8.3.1 Mechanical Properties

Mechanical properties are key to all engineering applications. In the case of marine infrastructure, they also play a crucial role. The provided research shows that geopolymer materials can have better properties in marine environments than traditional concrete. Zhang *et al.* [47] compared geopolymer concrete with traditional concrete. The materials were reinforced by using basalt bars. Geopolymer concrete was made from marine-based materials – seawater and coral aggregates [47]. The materials deterioration patterns and damage mechanisms under seawater immersion were investigated. The results show that geopolymer beams exhibited better resistance to seawater erosion than beams made from traditional concrete. After 12 months of exposition in seawater, the ultimate loading capacity of traditional concrete beams decreased by 12.2%, while that of geopolymer beams only degraded by 9.5% [47].

8.3.2 Durability and Resistance against Corrosive Environments

The topic of durability is presented in a number of studies on geopolymer materials in marine environments. It is one of the most important factors in the possible application of the material to this environment. Such investigation was provided by Luga *et al.* [48]. The main area of this investigation was the influence of wet-dry cycling in saline environments on the mechanical properties of geopolymer material. The results showed that geopolymer materials have significantly reduced water absorption and porosity with increasing wet–dry cycles compared to traditional concretes. Also, the mechanical properties of geopolymers were favorable. The findings confirmed the superior durability of geopolymer under harsh saline conditions. A durability investigation was also conducted on some reinforced geopolymers [49]. In this case, the steel reinforcements were replaced with epoxy-coated carbon yarn. The main point of the research was the resistance against chemical attacks, including acid, sulphate, marine water, and alkaline solutions. The results confirmed better durability of geopolymer material compared to cement concrete in terms of water absorption, acid resistance, sulphate resistance, marine water attack, and chloride diffusion [49]. The findings suggest that carbon yarn can be applied for special purposes in construction where corrosion resistance and durability are crucial [49].

Similar studies were made on geopolymer composites reinforced by basalt bars [50]. The materials were investigated under seawater drying-wetting cycles. The results show better resistance to seawater attacks on geopolymer concrete than traditional concrete. Moreover, the geopolymer composite reveals lower reductions in the bond strength with basalt reinforcement [50]. Another investigation on the durability of basalt fibre-reinforced geopolymer concrete takes into consideration corrosive ions attack (Cl^-, SO_4^{2-}, Mg^{2+}) in artificial seawater [51]. The results showed that the volume expansion rate and strength loss rate were more favorable for geopolymer concrete compared to traditional ones, especially the coefficient of ions diffusion was much lower in the case of geopolymer concrete [51]. It proves excellent resistance to ion migration of geopolymer materials and shows the possibility of application of this material in marine environments [51]. Rashad *et al.* investigated the improvement resistance of the geopolymer material by adding limestone powder [52]. The main aim of limestone powder addition to the geopolymer matrix was to prevent the deterioration caused by seawater attack and the simulated tidal. The results reveal that limestone powder in amounts up to 20% into geopolymer cement improved the compressive strength and limited the deterioration caused by the marine environment [52]. In turn, Kanagaraj *et al.* [53] compared corrosion resistance to three different concrete types: conventional cement concrete, self-compacting concrete, and geopolymer concrete [53]. According to their research, the best performance of long-term exposure to corrosive environments is due to geopolymer concrete [53]. Other authors confirmed these results. Sungkono *et al.* revealed that the permeability coefficient of geopolymer concrete is less than that of traditional concrete, while the corrosion resistance properties of geopolymer concrete indicated better results than traditional concrete. Moreover, the geopolymer concrete microstructure is more stable when corrosion occurs, which is a significant advantage for the marine infrastructure [54].

The important topic of investigation related to the marine environment is the resistance of the materials to chloride attack. The chloride environment is considered the main reason for the deterioration of the material properties in the environment. The performance of developed polyvinyl alcohol (PVA) fibre-reinforced geopolymer composites against the ingress of chloride ions was the main topic of research provided by Hasan and Hossain [55]. They investigated different types of raw materials used as a geopolymer matrix and noticed that the high calcium fly ash and slags have better resistance to chlorine attack and better chlorine durability than low calcium fly ash-based geopolymer composites. As an explanation of this phenomenon, they pointed to the higher density of the geopolymer matrix that is based on high calcium raw materials [55]. However, other research shows that slag addition is not always favorable for this kind of property [56]. According to this research, the slag addition showed a high rate of deterioration after marine exposure compared with the material without slag. It was caused by the reaction of NaCl salts with pores and flaws in the material, which strongly influenced the mechanical properties and durability performances of the specimens with slag [56].

The behavior of geopolymers under chloride attack was also the topic of the research provided by Somaiya [57]. The materials were tested with a chloride penetration test and curing with water consisting of H_2SO_4. The results showed good resistance of geopolymers to chloride ingress and the possibility of application of this material in marine environments [57].

Another important study was made on the influence of magnesium ion (Mg^{2+}) on geopolymers [58]. The study reveals that Mg^{2+} retards the setting time of geopolymers. These findings demonstrate the tailorability of setting time by the proper amount of Mg^{2+}. This can be achieved by using marine resources and using different geopolymer applications, including 3D printing technology [58].

8.3.3 Other Properties

An important area connected with applications in the marine environment in many areas is freeze-thaw behavior. The material behavior in changeable temperature could affect the safety and the lifetime of marine infrastructures [59]. This topic was investigated by Zhao *et al.* [59]. They studied the mechanical properties and frost durability of PVA fibre-reinforced geopolymer composites based on fly ash and bentonite. The research reveals a better influence of bentonite on frost properties [59].

8.4 APPLICATIONS IN MARINE ENVIRONMENT

The main application of geopolymer concrete is marine engineering construction, including the construction of ports, harbors, offshore platforms, and coastal protection structures [26, 60]. In the last period, coastal infrastructure is progressively becoming more vulnerable to climate change, including storms. Because of that, more and more advanced solutions are required for proper protection. One possible solution in this area can be more durable geopolymer materials [61]. The research suggests that one of the desired materials is fly ash, which has a superior synergistic influence on geopolymer concrete performance and very good resistance to chloride ingress and seawater corrosion [61]. This kind of solution was tested in practice as thirteen 16-tonne hangar units on the Port Kembla northern breakwater in Australia [61].

Li *et al.* also show that geopolymers can be applied to stabilizing marine soil using eco-friendly components, including industrial by-products. In this case, the main components of geopolymer materials were calcium carbide residue-activated coal gangue [62]. Other research in this area was done using coal gangue mixed with calcium carbide residues as raw material for geopolymer production [63]. The obtained material was successfully applied to enhance the geomechanical properties of marine clay [63]. Johney *et al.* confirmed these results and investigated the geopolymer binders as possible substitutes for lime. The main findings of this investigation confirm the usefulness of geopolymers for stabilizing marine clay deposits [64]. Geopolymers can also be used to stabilize marine sand. This kind of sand is characterized by uniformly grain-size distribution, which causes it to easily lose its shear strength in a loose condition and under cyclic loading [65]. Geopolymers can be applied as a stabilizing material to enhance the shear strength of marine sand [65]. Promising results have also been obtained in the case of marine clay stabilization [66].

Nowadays, the important application seems to be artificial coral reefs. This application requires form geopolymer materials that not only have excellent properties

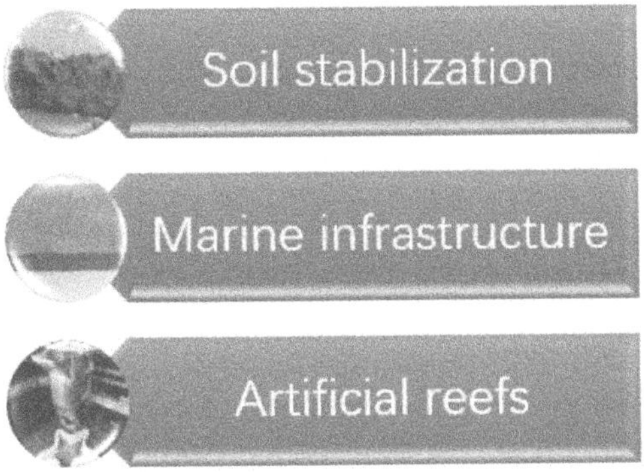

FIGURE 8.2 The main areas of geopolymer applications in the marine environment.

connected with behavior in marine environments but also most of them are produced using 3D printing technology. It also requires the specific properties connected with the additive manufacturing process [2, 67]. The product also has to fulfil environmental requirements [68]. This kind of research was provided by Santos *et al.* [69]. They provide ecotoxicological analysis of artificial reefs in the aquatic environment. They compare artificial reefs made by 3D printing based on cement and geopolymer mortars [69]. The provided test confirms the environmental acceptability of its potential use in both investigated materials; however, better results were obtained for traditional concrete [69]. Also, Ly *et al.* [70] compared different cement materials, such as geopolymers and cement concrete, assessing their durability and biofouling (colonization of structures by biological organisms) in seawater [70]. In this case, the results also indicate that cement is a better material for building an artificial reef using 3D printing due to its better mechanical properties and less susceptibility to biofouling [70]. In turn, the team of Yoris-Nobile *et al.* [68] showed that geopolymers have better biological receptivity compared to cement [68]. This indicates the need for further research on the selection of materials depending on the purpose of restoration (specific marine species) (see Figure 8.2).

8.5 CONCLUSIONS, CHALLENGES, AND FURTHER PERSPECTIVES

The provided analysis shows that geopolymers can be highly desirable materials for marine applications due to their physio-chemical properties, in particular, long-term stability and resistance to different environments. Currently, geopolymers are promising materials for marine applications, but their applications are connected to several challenges, and they still require development and optimization. This chapter recognizes the following areas as the most promising for future development:

1. The current research shows the possibility of the development of eco-friendly buildings and sustainable development in the island and coastal areas through the replacement of traditional ingredients of geopolymers with marine-based resources. This kind of component brings environmental benefits and usually has good corrosion resistance.

2. Basalt fibres and bars are used as the main tested reinforcement. They show better properties in marine environments than traditional steel bars. The other perspective areas are connected with polymeric and glass fibres.
3. The important research area is material properties, especially durability and resistance in marine environments. The research shows the possibility of using waste and industrial by-products to reinforce these properties in geopolymer compositions.
4. The application of geopolymers is focused on three areas: marine infrastructure, soil stabilization, and artificial reefs. All these areas required new solutions connected with material engineering.
5. Despite the widespread interest in these materials and intensive research, thus far, only a few prototype solutions have been developed and tested in relevant environments.

ACKNOWLEDGEMENTS

This research was supported by the project entitled "Development of geopolymer composites as a material for protection of hazardous wrecks and other critical underwater structures against corrosion" under the M-ERA.NET 3 program by the Polish National Centre for Research and Development, grant number M-ERA.NET3/2021/71/MAR-WRECK/2022.

REFERENCES

1. J. Gnanaraj S and Vasugi K., 'A comprehensive review of hydrophobic concrete: surface and bulk modifications for enhancing corrosion resistance', *Engineering Research Express*, vol. 6, no. 3, p. 032101, Sep. 2024, doi: 10.1088/2631-8695/ad5d55.
2. K. Korniejenko *et al.*, 'Additive manufacturing in underwater applications', *Applied Sciences*, vol. 14, no. 4, p. 1346, Feb. 2024, doi: 10.3390/app14041346.
3. F. Qu, W. Li, W. Dong, V. W. Y. Tam, and T. Yu, 'Durability deterioration of concrete under marine environment from material to structure: A critical review', *Journal of Building Engineering*, vol. 35, p. 102074, Mar. 2021, doi: 10.1016/j.jobe.2020.102074.
4. X. Chen, C. Yu, L. Wang, and B. Yu, 'A comprehensive review of the bio-corrosion mechanisms, hydrodynamics and antifouling measures on marine concrete', *Ocean Engineering*, vol. 310, p. 118696, Oct. 2024, doi: 10.1016/j.oceaneng.2024.118696.
5. G. Koch, 'Cost of corrosion', in *Trends in Oil and Gas Corrosion Research and Technologies*, Elsevier, 2017, pp. 3–30. doi: 10.1016/B978-0-08-101105-8.00001-2.
6. Th. Warscheid and J. Braams, 'Biodeterioration of stone: A review', *International Biodeterioration & Biodegradation*, vol. 46, no. 4, pp. 343–368, Dec. 2000, doi: 10.1016/S0964-8305(00)00109-8.
7. N. Ghafoori, M. Najimi, J. Sobhani, and M. A. Aqel, 'Predicting rapid chloride permeability of self-consolidating concrete: A comparative study on statistical and neural network models', *Construction and Building Materials*, vol. 44, pp. 381–390, Jul. 2013, doi: 10.1016/j.conbuildmat.2013.03.039.
8. N. Ganesan, R. Abraham, and S. Deepa Raj, 'Durability characteristics of steel fibre reinforced geopolymer concrete', *Construction and Building Materials*, vol. 93, pp. 471–476, Sep. 2015, doi: 10.1016/j.conbuildmat.2015.06.014.
9. M. Albitar, M. S. Mohamed Ali, P. Visintin, and M. Drechsler, 'Durability evaluation of geopolymer and conventional concretes', *Construction and Building Materials*, vol. 136, pp. 374–385, Apr. 2017, doi: 10.1016/j.conbuildmat.2017.01.056.

10. H. Rashidian-Dezfouli and P. R. Rangaraju, 'A comparative study on the durability of geopolymers produced with ground glass fiber, fly ash, and glass-powder in sodium sulfate solution', *Construction and Building Materials*, vol. 153, pp. 996–1009, Oct. 2017, doi: 10.1016/j.conbuildmat.2017.07.139.
11. Z. Zhang, X. Yao, and H. Zhu, 'Potential application of geopolymers as protection coatings for marine concrete. II. Microstructure and anticorrosion mechanism', *Applied Clay Science*, vol. 49, no. 1–2, pp. 7–12, Jun. 2010, doi: 10.1016/j.clay.2010.04.024.
12. G. Fahim Huseien, J. Mirza, M. Ismail, S. K. Ghoshal, and A. Abdulameer Hussein, 'Geopolymer mortars as sustainable repair material: A comprehensive review', *Renewable and Sustainable Energy Reviews*, vol. 80, pp. 54–74, Dec. 2017, doi: 10.1016/j.rser.2017.05.076.
13. J. R. Royer and D. D. Koo, 'Comparative analysis of geopolymer technology for sewer system rehabilitation', in *Pipelines 2015*, Baltimore, Maryland: American Society of Civil Engineers, Aug. 2015, pp. 1343–1354. doi: 10.1061/9780784479360.123.
14. J. R. Bone, R. Stafford, A. E. Hall, and R. J. H. Herbert, 'Biodeterioration and bioprotection of concrete assets in the coastal environment', *International Biodeterioration & Biodegradation*, vol. 175, p. 105507, Nov. 2022, doi: 10.1016/j.ibiod.2022.105507.
15. L. Jiang, T. R. Pettitt, N. Buenfeld, and S. R. Smith, 'A critical review of the physiological, ecological, physical and chemical factors influencing the microbial degradation of concrete by fungi', *Building and Environment*, vol. 214, p. 108925, Apr. 2022, doi: 10.1016/j.buildenv.2022.108925.
16. J. R. Bone, R. Stafford, A. E. Hall, and R. J. H. Herbert, 'The intrinsic primary bioreceptivity of concrete in the coastal environment – A review', *Developments in the Built Environment*, vol. 10, p. 100078, May 2022, doi: 10.1016/j.dibe.2022.100078.
17. T. Noeiaghaei, A. Mukherjee, N. Dhami, and S.-R. Chae, 'Biogenic deterioration of concrete and its mitigation technologies', *Construction and Building Materials*, vol. 149, pp. 575–586, Sep. 2017, doi: 10.1016/j.conbuildmat.2017.05.144.
18. G. Furtos, D. Prodan, C. Sarosi, D. Popa, M. Moldovan, and K. Korniejenko, 'The precursors used for developing geopolymer composites for circular economy—A review', *Materials*, vol. 17, no. 7, p. 1696, Apr. 2024, doi: 10.3390/ma17071696.
19. S. Pradhan, Z. Li, and S. Qian, 'A thermo-mechano-chemical activation technique to use quartz rich marine clay for one-part geopolymer preparation', *Cement and Concrete Composites*, vol. 140, p. 105057, Jul. 2023, doi: 10.1016/j.cemconcomp.2023.105057.
20. F. Todaro *et al.*, 'Recycling of contaminated marine sediment and industrial by-products through combined stabilization/solidification and granulation treatment', *Materials*, vol. 16, no. 6, p. 2399, Mar. 2023, doi: 10.3390/ma16062399.
21. L. Monteiro, J. Saliba, H. Yanez-Godoy, and N. Saiyouri, 'Strength and durability assessment of geopolymer mortars based on non-calcined dredged sediments', in *International RILEM Conference on Synergising Expertise towards Sustainability and Robustness of Cement-Based Materials and Concrete Structures*, vol. 44, A. Jędrzejewska, F. Kanavaris, M. Azenha, F. Benboudjema, and D. Schlicke, Eds., in RILEM Bookseries, vol. 44., Cham: Springer Nature Switzerland, 2023, pp. 385–394. doi: 10.1007/978-3-031-33187-9_36.
22. S. Irum and F. Shabbir, 'Performance of fly ash/GGBFS based geopolymer concrete with recycled fine and coarse aggregates at hot and ambient curing', *Journal of Building Engineering*, vol. 95, p. 110148, Oct. 2024, doi: 10.1016/j.jobe.2024.110148.
23. H. Yang, Q. Yang, J. Luo, J. Jiang, J. Mei, and A. Liu, 'Shear strength and failure criterion of geopolymer coral aggregate concrete under compression-shear loading', *Journal of Building Engineering*, vol. 76, p. 107241, Oct. 2023, doi: 10.1016/j.jobe.2023.107241.
24. M. Indhumathi Anbarasan, A. Leema Margret, V. Ragavan, and J. Ramprashath, 'Investigation on corrosion behaviour of geopolymer concrete using DMS and M–Sand as a fine aggregate under ambient curing conditions', *Materials Today: Proceedings*, p. S2214785323004789, Feb. 2023, doi: 10.1016/j.matpr.2023.01.383.

25. D. Xu, G. An, Y. Chen, Z. Liu, and X. Liu, 'Experiment on compressive properties and microscopic analysis of sea sand geopolymer-based recycled concrete', *Materials*, vol. 17, no. 1, p. 28, Dec. 2023, doi: 10.3390/ma17010028.
26. Y. Yang, S. Fang, W. Feng, S. Wan, L. Li, and Y. Tang, 'Flexural and compressive performance of BFRP-reinforced geopolymer sea-sand concrete beams and columns: Experimental and analytical investigation', *Composite Structures*, vol. 318, p. 117089, Aug. 2023, doi: 10.1016/j.compstruct.2023.117089.
27. Z. Yang, Z. Chen, H. Zhu, B. Zhang, Z. Dong, and X. Zhan, 'Efficient utilization of coral waste for internal curing material to prepare eco-friendly marine geopolymer concrete', *Journal of Environmental Management*, vol. 368, p. 122173, Sep. 2024, doi: 10.1016/j.jenvman.2024.122173.
28. B. Zhang, H. Peng, T. Xiong, and H. Zhu, 'Towards enhancing the durability of seawater coral aggregate concrete under drying-wetting cycles with slag-based geopolymers', *Journal of Sustainable Cement-Based Materials*, vol. 13, no. 3, pp. 389–401, Mar. 2024, doi: 10.1080/21650373.2023.2278846.
29. X. Fan, J. Zhu, and X. Gao, 'Sea/coral sand in marine engineered geopolymer composites: Engineering, mechanical, and microstructure properties', *International Journal of Applied Ceramic Technology*, paper no.14874, Jul. 2024, doi: 10.1111/ijac.14874.
30. M. Saba, F. F. Tehrani, P. Hajikarimi, and J. Absi, 'Investigation on partially replacing metakaolin with marine shell waste to produce sustainable eco-friendly geopolymer mortars', *Clean Technologies and Environmental Policy*, vol. 25, no. 8, pp. 2639–2653, Oct. 2023, doi: 10.1007/s10098-023-02512-4.
31. A. Tahsin and W. Ashraf, 'Unveiling the carbon footprint reduction potential of cementitious composites: CO_2 sequestration in a lime-clay binder incorporating medium-grade clay', *ACS Sustainable Chemistry & Engineering*, vol. 12, no. 30, pp. 11099–11110, Jul. 2024, doi: 10.1021/acssuschemeng.3c08555.
32. Z. Li *et al.*, 'Seawater used to metakaolinite-based geopolymer preparation', *Construction and Building Materials*, vol. 392, p. 131816, Aug. 2023, doi: 10.1016/j.conbuildmat.2023.131816.
33. Z. Yang *et al.*, 'Synthesis of eco-sustainable seawater sea-sand geopolymer mortars from ternary solid waste: Influence of microstructure evolution on mechanical performance', *Sustainable Materials and Technologies*, vol. 41, p. e01056, Sep. 2024, doi: 10.1016/j.susmat.2024.e01056.
34. J.-C. Lao, B.-T. Huang, L.-Y. Xu, M. Khan, Y. Fang, and J.-G. Dai, 'Seawater sea-sand engineered geopolymer composites (EGC) with high strength and high ductility', *Cement and Concrete Composites*, vol. 138, p. 104998, Apr. 2023, doi: 10.1016/j.cemconcomp.2023.104998.
35. M. M. Calvino, G. Cavallaro, S. Milioto, and G. Lazzara, 'Composite materials based on halloysite clay nanotubes and cellulose from *Posidonia oceanica* sea balls: From films to geopolymers', *Environmental Science: Nano*, vol. 11, no. 4, pp. 1508–1520, 2024, doi: 10.1039/D3EN00879G.
36. J. Wei, C. Liu, J. Liu, X. Yu, S. Xu, and Y. Su, 'Investigations on geopolymer-based seawater sea-sand high performance concrete slabs reinforced with basalt FRP bars under direct contact explosions', *Construction and Building Materials*, vol. 411, p. 134538, Jan. 2024, doi: 10.1016/j.conbuildmat.2023.134538.
37. Z. Lu, C. Zhao, J. Zhao, C. Shi, and J. Xie, 'Bond durability of FRP bars and seawater–sea sand–geopolymer concrete: Coupled effects of seawater immersion and sustained load', *Construction and Building Materials*, vol. 400, p. 132667, Oct. 2023, doi: 10.1016/j.conbuildmat.2023.132667.
38. Z. Yang *et al.*, 'Eco-sustainable design of seawater sea-sand slag-based geopolymer mortars incorporating ternary solid waste', *Construction and Building Materials*, vol. 431, p. 136512, Jun. 2024, doi: 10.1016/j.conbuildmat.2024.136512.

39. Y. H. Zhang, W. L. Zhong, and L. F. Fan, ‘Long-term durability investigation of basalt fiber-reinforced geopolymer concrete in marine environment’, *Journal of Materials Research and Technology*, vol. 31, pp. 593–605, Jul. 2024, doi: 10.1016/j.jmrt.2024.06.078.
40. S. K. Rahman and R. Al-Ameri, ‘Long-term performance of basalt fibre-reinforced marine geopolymer concrete in harsh environment’, *Magazine of Concrete Research*, vol. 75, no. 22, pp. 1165–1187, Nov. 2023, doi: 10.1680/jmacr.23.00035.
41. J. Zhao *et al.*, ‘Coupling effects of seawater immersion and prestressing on the durability of BFRP bars embedded in seawater–sea sand geopolymer mortars’, *Journal of Materials in Civil Engineering*, vol. 36, no. 4, p. 04024049, Apr. 2024, doi: 10.1061/JMCEE7.MTENG-17106.
42. D. Huang, Z. Liu, Y. Lu, W. Ma, and S. Li, ‘Experimental and numerical investigation of bias performance of steel fiber-reinforced GRC-CFFT columns’, *Engineering Structures*, vol. 319, p. 118897, Nov. 2024, doi: 10.1016/j.engstruct.2024.118897.
43. R. Mohana and S. M. L. Bharathi, ‘Parametric investigation on the novel and cost-effective nano fly ash impregnated geopolymer system for sustainable construction’, *Frontiers of Structural and Civil Engineering*, vol. 18, no. 2, pp. 170–183, Feb. 2024, doi: 10.1007/s11709-024-1010-5.
44. W. L. Zhong, Y. H. Zhang, L. F. Fan, and P. F. Li, ‘Effect of PDMS content on waterproofing and mechanical properties of geopolymer composites’, *Ceramics International*, vol. 48, no. 18, pp. 26248–26257, Sep. 2022, doi: 10.1016/j.ceramint.2022.05.306.
45. M. Sun, Y. Qin, J. Tan, J. Liu, J. Li, and X. Cui, ‘Doping and superhydrophobic modification for improving marine antifouling performance of alkali-based geopolymer coating’, *Coatings*, vol. 14, no. 8, p. 974, Aug. 2024, doi: 10.3390/coatings14080974.
46. C. Phiangphimai *et al.*, ‘Durability properties of novel coating material produced by alkali-activated/cement powder’, *Construction and Building Materials*, vol. 363, p. 129837, Jan. 2023, doi: 10.1016/j.conbuildmat.2022.129837.
47. B. Zhang, H. Zhu, C. You, Z. Yang, J. Liu, and H. Peng, ‘Flexural durability of BFRP bars reinforced geopolymer-based coral aggregate concrete beams conditioned in marine environments’, *Journal of Building Engineering*, vol. 94, p. 109959, Oct. 2024, doi: 10.1016/j.jobe.2024.109959.
48. E. Luga, E. Mustafaraj, M. Corradi, and C. D. Atiş, ‘Alkali-activated binders as sustainable alternatives to Portland cement and their resistance to saline water’, *Materials*, vol. 17, no. 17, p. 4408, Sep. 2024, doi: 10.3390/ma17174408.
49. A. E. Alexander and A. P. Shashikala, ‘Studies on the mechanical and durability performance of textile reinforced geopolymer concrete beams’, *Materials Today Communications*, vol. 35, p. 105837, Jun. 2023, doi: 10.1016/j.mtcomm.2023.105837.
50. B. Zhang, F. Xu, H. Zhu, Z. Yang, and H. Peng, ‘Deterioration of bond performance between BFRP bars and coral aggregate concrete incorporating slag-based geopolymers under seawater corrosion environments’, *Construction and Building Materials*, vol. 411, p. 134518, Jan. 2024, doi: 10.1016/j.conbuildmat.2023.134518.
51. Z. Chen, J. Yu, Y. Nong, Y. Yang, H. Zhang, and Y. Tang, ‘Beyond time: Enhancing corrosion resistance of geopolymer concrete and BFRP bars in seawater’, *Composite Structures*, vol. 322, p. 117439, Oct. 2023, doi: 10.1016/j.compstruct.2023.117439.
52. A. M. Rashad, M. Ezzat, A. M. ElNagar, and M. H. El-Nashar, ‘Valorization of limestone powder as an additive for fly ash geopolymer cement under the effect of the simulated tidal zone and seawater attack’, *Construction and Building Materials*, vol. 369, p. 130616, Mar. 2023, doi: 10.1016/j.conbuildmat.2023.130616.
53. B. Kanagaraj, R. Priyanka, N. Anand, T. Kiran, A. D. Andrushia, and E. Lubloy, ‘A sustainable solution for mitigating environmental corrosion in the construction sector and its socio-economic concern’, *Case Studies in Construction Materials*, vol. 20, p. e03089, Jul. 2024, doi: 10.1016/j.cscm.2024.e03089.

54. K. K. D. Sungkono, I. Satyarno, H. Priyosulistyo, and I. Perdana, 'Corrosion resistance of high calcium fly ash based reinforced geopolymer concrete in marine environment', *Civil Engineering and Architecture*, vol. 11, no. 5A, pp. 3175–3189, Sep. 2023, doi: 10.13189/cea.2023.110827.
55. M. J. Hasan and K. M. A. Hossain, 'Assessing suitability of geopolymer composites under chloride exposure', in *Proceedings of the Canadian Society of Civil Engineering Annual Conference 2021*, vol. 240, S. Walbridge, M. Nik-Bakht, K. T. W. Ng, M. Shome, M. S. Alam, A. El Damatty, and G. Lovegrove, Eds., in Lecture Notes in Civil Engineering, vol. 240, Singapore: Springer Nature Singapore, 2023, pp. 375–387. doi: 10.1007/978-981-19-0507-0_35.
56. C. R. Gadikota and D. S. Chandra, 'Influence of slag content on mechanical, durability and microstructural performance and their correlations in fly ash geopolymers exposed to marine environment', *Asian Journal of Civil Engineering*, vol. 25, no. 2, pp. 2091–2108, Feb. 2024, doi: 10.1007/s42107-023-00896-8.
57. P. Somaiya, 'Development of geopolymer concrete using waste marble sand in marine environments', in *2023 IEEE 11th Region 10 Humanitarian Technology Conference (R10-HTC)*, Rajkot, India: IEEE, Oct. 2023, pp. 1016–1021. doi: 10.1109/R10-HTC57504.2023.10461775.
58. S. S. Zhang, S. Wang, and X. Chen, 'Understanding the role of magnesium ions on setting of metakaolin-based geopolymer', *Cement and Concrete Research*, vol. 177, p. 107430, Mar. 2024, doi: 10.1016/j.cemconres.2024.107430.
59. N. Zhao *et al.*, 'Effects of fly ash and bentonite on mechanical and durability properties of fiber reinforced geopolymer', *JOM*, vol. 75, no. 3, pp. 848–858, Mar. 2023, doi: 10.1007/s11837-022-05665-7.
60. B. Gopalakrishna and P. Dinakar, 'An innovative approach to fly ash-based geopolymer concrete mix design: Utilizing 100% recycled aggregates', *Structures*, vol. 66, p. 106819, Aug. 2024, doi: 10.1016/j.istruc.2024.106819.
61. Engineers Australia, Ed., *Australasian Coasts & Ports 2023: Twin Waters, Australia, 15-18 April 2023.* Red Hook, NY: Curran Associates, Inc, 2024.
62. J. Li, Y. Shan, P. Ni, J. Cui, Y. Li, and J. Zhou, 'Mechanics, durability, and microstructure analysis of marine soil stabilized by an eco-friendly calcium carbide residue-activated coal gangue geopolymer', *Case Studies in Construction Materials*, vol. 20, p. e02687, Jul. 2024, doi: 10.1016/j.cscm.2023.e02687.
63. J. Li, Y. Shan, P. Ni, Y. Li, J. Cui, and J. Zhou, 'Multiscale experimental analysis of marine clay stabilized with coal gangue–calcium carbide residue geopolymer', *Acta Geotechnica*, vol. 18, no. 11, pp. 5921–5939, Nov. 2023, doi: 10.1007/s11440-023-02055-4.
64. A. Johney, B. M. Abraham, and D. K. Sahoo, 'Metakaolin/Slag based Geopolymers – An alternative to lime stabilization of Cochin Marine clays', *International Journal of Geotechnical Engineering*, vol. 18, no. 3, pp. 262–274, Mar. 2024, doi: 10.1080/19386362.2024.2388590.
65. A. S. Muntohar, M. Afzalurrahman, and W. Diana, 'The undrained-unconsolidated shear strength behavior of the geopolymer-stabilized marine sand', presented at the XVII Mexican Symposium on Medical Physics, Veracruz, México, 2023, p. 050009. doi: 10.1063/5.0154401.
66. F. Pakir *et al.*, 'Microstructure behaviour of the marine clay treated with geopolymer', *IOP Conference Series: Earth and Environmental Science*, vol. 1249, no. 1, p. 012042, Oct. 2023, doi: 10.1088/1755-1315/1249/1/012042.
67. I. V. Matus, J. L. Alves, J. Góis, P. Vaz-Pires, and A. Barata Da Rocha, 'Artificial reefs through additive manufacturing: A review of their design, purposes and fabrication process for marine restoration and management', *Rapid Prototyping Journal*, vol. 30, no. 11, pp. 87–122, Apr. 2024, doi: 10.1108/RPJ-07-2023-0222.

68. A. I. Yoris-Nobile *et al.*, 'Artificial reefs built by 3D printing: Systematisation in the design, material selection and fabrication', *Construction and Building Materials*, vol. 362, p. 129766, Jan. 2023, doi: 10.1016/j.conbuildmat.2022.129766.
69. J. Santos *et al.*, 'Assessment of the environmental acceptability of potential artificial reef materials using two ecotoxicity tests: Luminescent bacteria and sea urchin embryogenesis', *Chemosphere*, vol. 310, p. 136773, Jan. 2023, doi: 10.1016/j.chemosphere.2022.136773.
70. O. Ly *et al.*, 'Optimisation of 3D printed concrete for artificial reefs: Biofouling and mechanical analysis', *Construction and Building Materials*, vol. 272, p. 121649, Feb. 2021, doi: 10.1016/j.conbuildmat.2020.121649.

9 Geopolymer Cement in Fixation of Industrial Waste Phosphoric Acid

Kaibao Wang, Qingxin Wei, Hongwei Chen, and Huirong Le

9.1 INTRODUCTION

With the rapid development of industrial modernisation and economic level worldwide, an excessive amount of waste acids is produced. Industrial waste acid has high acidity and often contains heavy metals, which may cause significant harm to the environment and lives. Waste acid could significantly affect the quality of surface water, threaten the survival of aquatic organisms, and damage plant growth due to the change in soil acidity and alkalinity. Additionally, heavy metal elements can also damage human organs seriously and cause a variety of diseases [1]. The high-quality development strategic goals, such as carbon neutrality, industrial waste and heavy metal pollution treatment, and green development, have been introduced successively; how to deal with heavy metal waste more economically and effectively is an important issue to be solved. Therefore, it is crucial to develop low-carbon materials and realise efficient solidification of heavy metal ions.

Heavy metal solidification refers to the conversion of heavy metals into environmentally acceptable wastes by physical or chemical mixing with adhesives for land disposal or construction purposes. This can be achieved via cement, polymer, and glass solidification [2]. Currently, cement solidification technology has been widely used to treat toxic and harmful wastes [3]. However, the drawbacks of cement-based solidification technology, such as high leaching rate, poor fixing effect, and heavy metals in waste, significantly affect the setting time and strength of cement [2]. In addition, cement-based materials emit excessive carbon dioxide during production. The solidification effect of glass-based materials is moderate, but the obtained products can only be landfilled or used as aggregate, which has a low added value [4].

Geopolymer is a three-dimensional inorganic polymer that is derived from the reaction between an aluminosilicate source and alkaline/acidic activators [5]. It exhibits excellent thermal stability, relatively high strength, and fire and chemical resistance. Besides, geopolymer has lower energy consumption and carbon emissions compared to traditional cement. Therefore, geopolymer has a promising potential for application in infrastructure, sewage treatment, and heavy metal solidification [4, 6]. The heavy metal ions are mainly solidified through physical (including physical storage

 DOI: 10.1201/9781003535683-9

and adsorption) and chemical methods [3]. The heavy metal solidification of alkaline-activated geopolymer has been widely studied [7–11]. Solid wastes containing heavy metals are used to prepare geopolymers under the activation of alkali activators, which can realise the reuse of solid waste and the safe solidification of heavy metal elements [4]. In this way, the heavy metal ions are solidified via a physical storage process.

In recent years, the acid-based geopolymer has shown great interest due to excellent temperature stability and mechanical strength [12–16]. In addition, compared with the alkali activation solution, the acid activation solution is more sustainable and environmentally friendly regarding greenhouse gas emissions and energy consumption [17]. Figure 9.1 shows the activation process of acidic- and alkaline-based geopolymers [17]. Pu et al. [13] found that phosphate-based geopolymer has a good fixing effect on lead metal, and its performance is better than that of alkali-activated geopolymer and ordinary Portland cement (OPC), especially in an acidic environment. Liu et al. [18] prepared a porous metakaolin-based geopolymer and discussed its application in wastewater treatment. The results showed that the porous structure of geopolymer

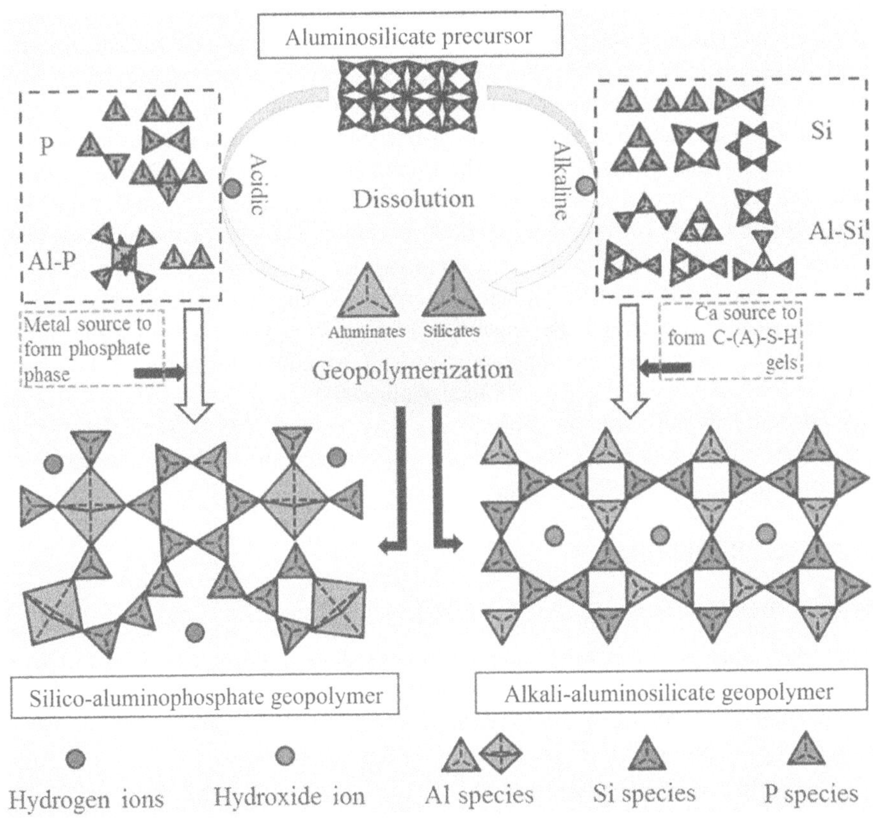

FIGURE 9.1 Activation process of aluminosilicate sources under acidic and alkaline conditions to form geopolymer matrices. (Reprinted with permission from ref. [17]. Copyright 2019, Frontiers and ref. [20]. Copyright 2024, Elsevier.)

could effectively adsorb heavy metals such as lead, nickel, and cadmium in wastewater, and the adsorption performance of acid-activated geopolymer was better than that of alkali-activated geopolymer at similar metal concentrations. He et al. [19] developed a low-cost phosphate acid-activated geopolymer; however, the compressive strength is relatively low (0.52–13.23 MPa) due to a low phosphate acid concentration, and the heavy metal solidification characteristics were not discussed.

Although the heavy metal solidification of acid-based geopolymer has been explored [14–16], the use of waste phosphoric acid as the activator is rarely mentioned. The gradual shortage of phosphate resources and the high price of phosphoric acid will be the key factors limiting the broad application of phosphoric acid-activated geopolymers in the future. Therefore, it is of great significance to expand the range and types of acid activators, such as the application of waste liquid containing phosphoric acid.

According to the survey conducted by the Qianzhan Research Institute in 2020 [21], the annual output of industrial waste acid in China reaches 100 million tons, of which no more than 40% is recycled. About 40% of the waste acid is directly neutralised; the rest is either turned into waste acid inventory reserves or unregulated storage. The estimated waste acid reserves exceed 200 million tons. Industrial waste acid has high acidity and often contains heavy metals and organic matter, which may cause significant harm to the environment. Neutralisation is the primary treatment method for waste acid in China, but its resource utilisation rate is insufficient. The gas generation and diffusion during the treatment process cause secondary pollution. The remaining salt residue and heavy metal salt treatment is complex, and the treatment process has a specific technical threshold. The proportion of waste acid recovery is constantly increasing, but there is a lack of demonstration projects for reference, and small and medium-sized enterprises occupy a large proportion [21].

This chapter discovered the feasibility of preparing geopolymers using waste phosphate acid with environmental and economic benefits. The novelty of this chapter is to provide a key technology for the harmless treatment of industrial waste acid and the high cost and resource waste in waste acid treatment. At the same time, it can reduce the pollution of heavy metals to the environment and the carbon emission of the construction and ceramic industry.

9.2 EXPERIMENTAL

9.2.1 Material

The waste acids A and B were supplied by Grandblue Environment Co., Ltd with unknown chemical composition. Thus, it is compulsory to conduct the chemical composition test, and this is described in Section 9.2.2.1. The kaolin and phosphorous acid (85% in water) were purchased from Sigma Aldrich.

9.2.2 Sample Preparation

9.2.2.1 Chemical Composition Determination of Waste Acid

Since the phosphate element concentration of the waste acid was excessive to allow ICP tests to be performed directly, and, if diluted, other low-concentration elements may not be measurable, calcium oxide (CaO) was first introduced for sedimentation, and the chemical composition of the filtered solution and sediment was then measured separately. The chemical composition was finally calculated by subtracting the mass of CaO. The detailed procedure is described as follows:

1. The mass of waste acids A and B was set to 50 g. The CaO was added into the solution slowly and continuously stirred until a pH value of 8 was obtained. The amount of CaO added was recorded.
2. The mixed solution was then filtered, and the sediment was weighed after drying. The filtered solution was set to 500 ml.
3. The chemical content of sediment was characterised via XRF, and the filtered solution was determined by the ICP test. Six metal elements that were relatively high in content were selected.
4. Since the waste acid emits a strong volatile odour, the waste acids A and B were diluted to conduct the Headspace Gas Chromatography test.
5. The chemical composition of the waste acids A and B was finally calculated by subtracting the mass of CaO.

9.2.2.2 Fabrication of Waste Acid-Activated Geopolymer

The kaolin (chemical composition shown in Table 9.1) was calcined at 800°C for 2 h to obtain Metakaolin (MK). The solid-liquid (waste acid) ratio was set to 1:1. Pure phosphoric acid or deionised water was then added to the slurry to obtain a P/Al molar ratio of 1.3:1. The slurry was continuously stirred for 30 min at 300 rpm. The mould was then placed on a shaker for 5 min, allowing gas bubbles to escape from the slurry before transferring into an oven at a temperature of 60°C for 24 h. Finally, the sample was de-moulded and conditioned in the ambient environment for seven days before mechanical measurement. The fabrication process of waste phosphate acid-activated geopolymer is illustrated in Figure 9.2. For simplicity, the geopolymer activated with waste acids A and B is named WPAG-A and WPAG-B, respectively.

9.2.3 Sample Characterisation

9.2.3.1 Chemical Composition and Crystalline Structure

To determine the crystalline phases of the specimens, X-ray diffraction (XRD) tests were carried out (D8 Advance, Bruker Optics, Ettlingen, Germany) with a scanning range of 10° to 80°. X-ray fluorescence spectrometry (XRF; ARL PERFORM X;

TABLE 9.1
Chemical Composition of Metakaolin

SiO_2	Al_2O_3	K_2O	Fe_2O_3	TiO_2	PbO	P_2O_5	CaO
54.37	39.78	2.11	0.916	0.662	0.602	0.445	0.186

Source: Reprinted with permission from ref. [20]. Copyright 2024, Elsevier.

FIGURE 9.2 The fabrication process of waste phosphate acid-activated geopolymer. (Reprinted with permission from ref. [20]. Copyright 2024, Elsevier.)

Thermo Fisher; Shanghai, China) was used to determine the chemical composition of the powder sample.

9.2.3.2 SEM Imaging

The morphology of geopolymer was obtained using a scanning electron microscope (SEM; Merlin, Zeiss Company, Oberkochen, Germany) with 15 kV voltage on the fractal surface of samples to discover the influence of acid content on the structure. The samples were first coated with platinum using Leica EM ACE600, and the coating thickness was set to 8 nm.

9.2.3.3 Infiltration Test

The infiltration test was performed following the HT 557-2010 standard for solid waste extraction procedures for leaching toxicity using the horizontal vibration method with X-ray fluorescence spectrometry, and the HJ 766-2015 standard for determining metals in solid waste using inductively coupled plasma mass spectrometry (ICP-MS) (Thermo, IRIS Intrepid II, China). The prepared geopolymer sample was vacuum dried at 105°C for 4 h, and it was then crushed and sieved with a 60 mesh sieve. An extraction agent (in this case water) was mixed with the crushed geopolymer powder at a solid-liquid ratio of 1:10 (i.e., 50 g geopolymer and 500 g water). The mixture was then placed on a shaker at 120 rpm for 8 h, followed by standing for 16 h to settle. The liquid was finally extracted by filtering the mixture, and an ICP test was performed to measure the concentration of the metal ions leached. The heavy metal solidification efficiency of the geopolymer was calculated based on the mass of the metal element leached into the solution divided by the total mass of the metal element in the waste acid.

9.2.3.4 Mechanical Test

The compression strength measurement was carried out using a universal mechanical test machine (Tinius Olsen 50ST) with a loading rate of 1 mm/min according to test standard GB/T 17671-2021. An extensometer was attached to the machine to determine the relative change in the gauge length on the test specimen at any time during the test. The change was automatically recorded and measured with an accuracy of 1% of the relevant value. The sample dimension for the compression test was 20 mm × 20 mm × 20 mm cube. The flexural strength was measured with a loading rate of 0.5 mm/min. The sample dimensions for the flexural test were 20 mm × 20 mm × 80 mm, with a span distance of 50 mm. A micrometre was used to measure the reading to 0.01 mm and provided with means for measuring the thickness and width of the test specimens. A minimum of five specimens were tested to calculate an average strength. The arithmetic means of the test results, the standard deviations, and 95% confidence intervals of the mean values are needed to be calculated.

9.2.3.5 Thermal Gravimetric Analysis

Thermal stability was tested by thermal gravimetric analysis (TGA, PerkinElmer STA8000), according to GB/T 13464 thermal analysis test methods for thermal stability of materials. The samples were heat-treated in the nitrogen from 30°C up to 1500°C with a rate of 20°C/min.

9.3 RESULTS AND DISCUSSION

9.3.1 Morphology and Crystalline Structure

Figure 9.3 shows the digital (a and b), optical (c and d), and SEM images (e and f) of WPAG-A and WPAG-B, respectively. In the macroscopic view, the digital and optical images clearly indicate a consolidated structure for WPAG-A. In contrast, the geopolymer activated with waste acid B shows a loose and porous structure. In the microscopic view, the geopolymer activated with waste acid A exhibits a lamellar structure, as indicated in Figure 9.3e, where amorphous glass geopolymer matrices with clear particles were observed in Figure 9.3f. These particles were most likely the quartz crystals, as the quartz from Metakaolin did not dissolve fully in the gel phase, meaning that a low degree of polymerisation reaction was performed [22]. This can be visualised in Figure 9.5 (XRD). Besides, since the waste acid B contains sulphuric acid (Table 9.2), metal elements, such as Mg, Al, and Fe, may react with H_2SO_4, resulting in gas (H_2) being generated. The gas bubbles are trapped within the slurry, expanding and generating mostly macropores. During the gas generation process, the ultimate shear stress of aerated slurry rises instantaneously [23]. Thus,

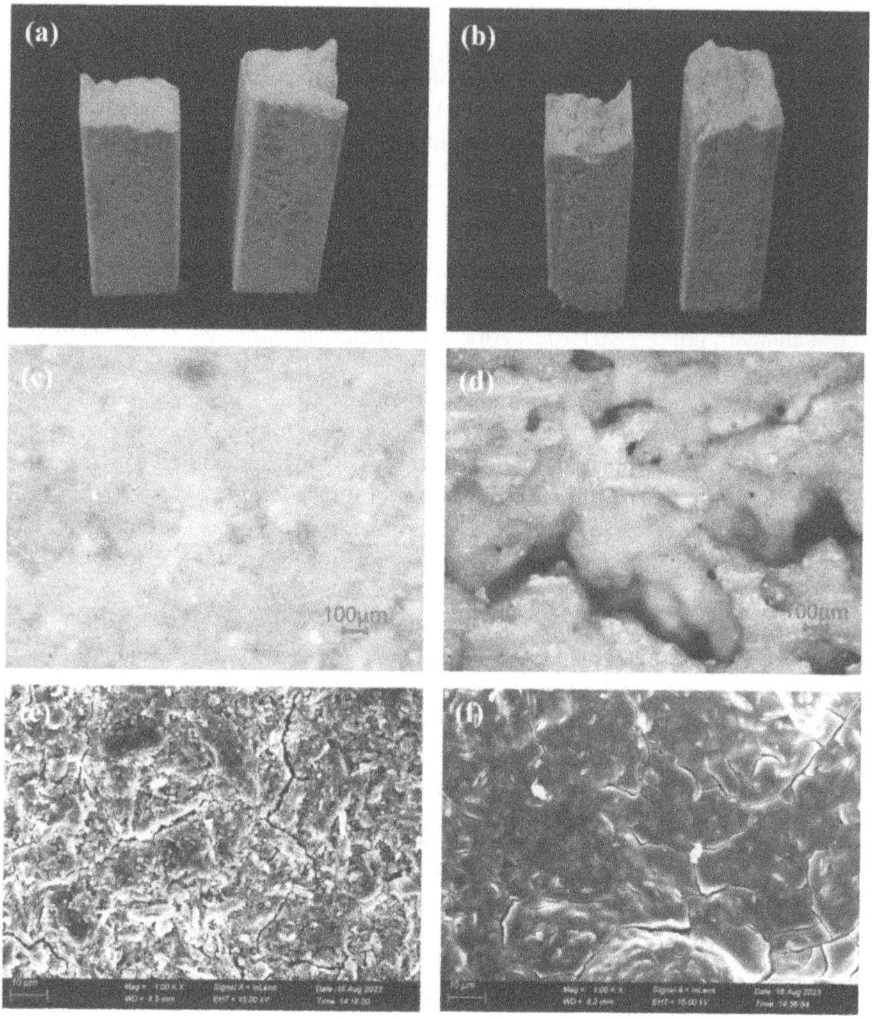

FIGURE 9.3 Digital images (a and b), optical images ×20 (c and d), and SEM images ×1000 (e and f) of WPAG-A and WPAG-B, respectively. (Reprinted with permission from ref. [20]. Copyright 2024, Elsevier.)

the generated bubbles are merged into larger pores in a small range. Therefore, the geopolymer activated with waste acid B shows a loose and porous structure.

Figure 9.4 shows the EDX analysis of WPAG-A and WPAG-B. Spectrum 3 indicates that the specimens are mainly composed of silicon, oxygen, phosphate, and aluminium with a limited amount of sulphur, whereas spectrum 4 for waste acid B clearly shows a significantly higher amount of sulphur presented within the specimen. In spectrum 3 (waste acid A), the Al atom percent is 8.74%, the Si atom percent is 10.02%, the P atom percent is 10.81%, and the S atom percent is 0.35%. In spectrum 4 (waste acid B), the Al atom percent is 5.26%, the Si atom percent is 4.38%,

TABLE 9.2
Chemical Composition of Waste Acid Sediment via XRF

	Content (wt.%)										
Element	**Ca**	**Px**	**Mo**	**Al**	**Mg**	**Na**	**Si**	**Fe**	**Sx**	**K**	**Ag**
Waste acid A	**43.23**	**16.1**	0.695	0.207	0.203	0.227	0.101	0.063	0.0309	0.062	0.0278
	Sr	Ti	Nd	W	Cu	Ni	Mn	Cl	Au	Ar	
	0.0203	0.0096	0.0135	0.0076	0.0065	0.0049	0.0033	/	/	/	
Waste acid B	Ca	Px	Mo	Al	Mg	Na	Si	Fe	Sx	K	Ag
	51.57	**6.28**	/	0.188	0.28	0.232	0.0964	0.0605	**4.73**	0.0144	0.0311
	Sr	Ti	Nd	W	Cu	Ni	Mn	Cl	Au	Ar	
	0.0315	0.0656	/	0.0113	0.0047	0.0038	0.0046	0.0105	0.0073	0.0021	

Source: Reprinted with permission from ref. [20]. Copyright 2024, Elsevier.

the P atom percent is 5.21%, and the S atom percent is 1.76%. The presence of silicon and phosphate confirms the formation of -Si-O-P-O-Si- networks, where the one of aluminium suggests the formation of new crystalline phases such as berlinite or aluminium phosphate ($AlPO_4$) [24].

Figure 9.5 shows the XRD patterns of WPAG-A and WPAG-B. Both patterns show a diffuse peak between 15 and 30° (2θ), indicating the characteristics of the

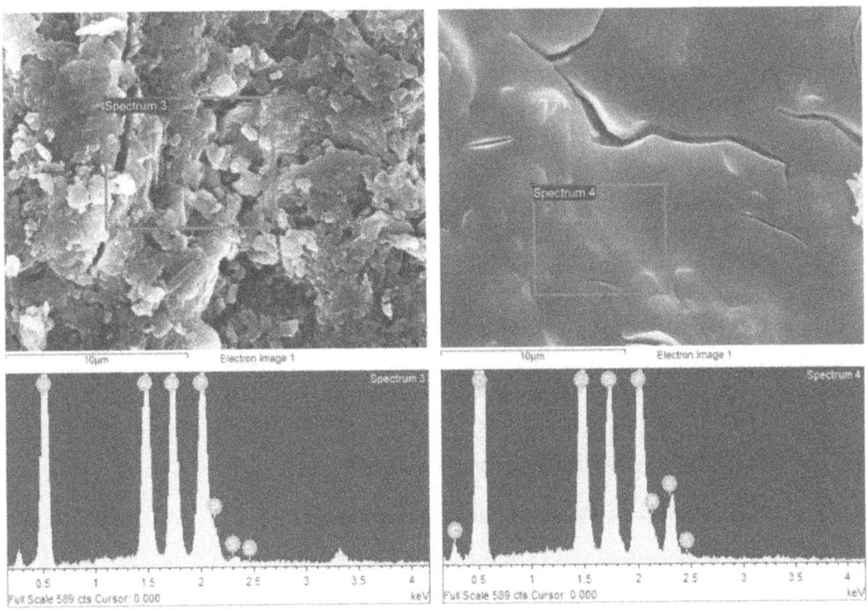

FIGURE 9.4 EDX analysis of WPAG-A and WPAG-B. (Reprinted with permission from ref. [20]. Copyright 2024, Elsevier.)

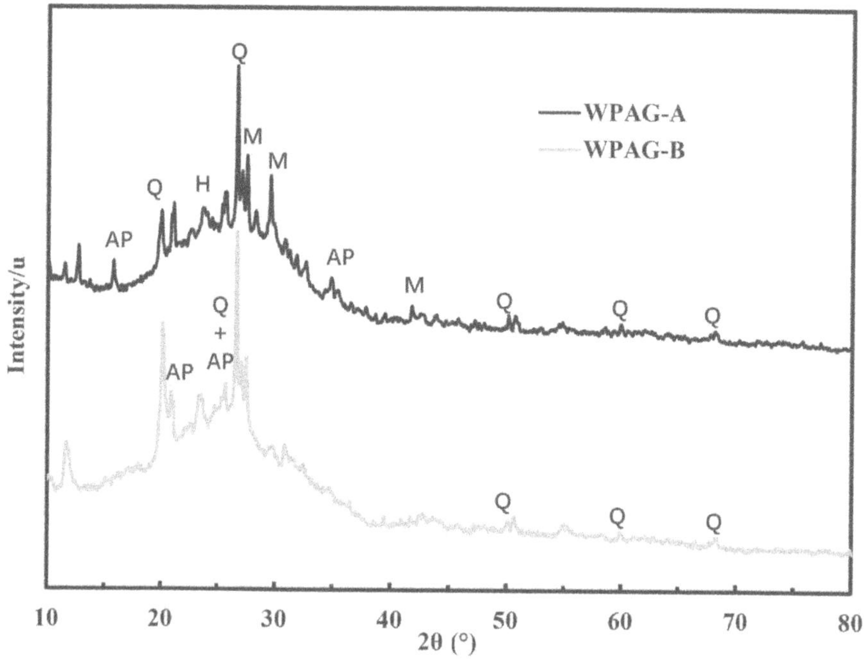

FIGURE 9.5 XRD patterns of WPAG-A and WPAG-B (Q-Quartz, SiO_2; M-Mullite; AP-Aluminum phosphate or berlinite; H-Hematite). (Reprinted with permission from ref. [20]. Copyright 2024, Elsevier.)

amorphous aluminosilicate gels of the reaction products in both samples. The predominant crystal phases in both samples were quartz (Q, SiO_2) and mullite (M, $3Al_2O_3 \cdot 2SiO_2$), and a few new crystalline phases, including aluminium phosphate (AP, $AlPO_4$) and monetite ($CaHPO_4$), were also observed, as indicated between 20 and 28° (2θ). The remaining quartz peak indicated that excessive quartz from Metakaolin did not take part in the chemical reaction [22]. It is worth mentioning that a higher Q peak and less amount of AP peaks were observed in waste acid B compared to waste acid A, and this may suggest that a high concentration of sulphuric acid inhibited the dissolution of MK and subsequent polymerisation reaction [19, 25]. As a result, a higher degree of polymerisation reaction leads to a consolidated structure of the geopolymer activated with waste acid A.

9.3.2 Heavy Metal Solidification Efficiency

9.3.2.1 Calculation of Waste Acid Concentration

Table 9.2 shows the chemical composition of waste acid sediment. 24.7 g of CaO was consumed for waste acid A to reach a pH of 8, while 47.3 g of CaO was consumed for waste acid B. The mass of waste acid sediments A and B weighed 45 and 85 g, respectively. The sediments of waste acid A mainly contain calcium and phosphate, and the weight percentage of the Px element is 16.1%, which is equal

TABLE 9.3
Chemical Composition of the Filtered Solution via ICP (Diluted to 500 ml)

	μg/ml					
Element	**Ag**	**Cu**	**Fe**	**Mn**	**Mo**	**Ni**
Waste acid A	0.025	2.420	0.456	0.431	1.297	0.130
Waste acid B	0.028	4.912	0.408	0.777	0.198	0.857

Source: Reprinted with permission from ref. [20]. Copyright 2024, Elsevier.

to 7.245 g (45 g × 16.1%, 0.234 mol). Since there is no phosphate or a tiny portion in the filtered solution, as indicated in Table 9.3, it is assumed that all phosphate comes from waste acid. If all the phosphorus is used as phosphoric acid (0.234 mol), there is 22.932 g of phosphoric acid in 50 g of waste acid, and the mass fraction is about 45.8%. Moreover, waste acid A also contains a moderate percentage of Mo and Al elements. Since these elements are insoluble in water, they can only exist in the form of acid. The slightly green colour of waste acid A also confirms the existence of molybdate. There are no other large number of positive ions coordinating with phosphate, so it was suggested that the phosphate can only exist in the form of phosphoric acid. The Headspace Gas Chromatography test confirms that the volatile substance in waste acid A is acetic acid; however, the content is unknown.

In contrast, the sediments of waste acid B mainly contain calcium, phosphate, and sulphur elements. Since the number of positive metal ions that can coordinate with sulphate is insufficient, it is assumed that all phosphate and sulphur elements in the sediments can be converted to phosphate and sulphuric acid. The weight of the Px element is calculated to be 5.338 g (6.28% ×85 g, 0.172 mol), which is equivalent to 16.89 g of phosphate acid in 50 g of waste acid B. Similarly, the weight of Sx element is calculated to be 4.021 g (4.73% ×85 g, 0.125 mol), which is equivalent to 12.30 g of sulphuric acid in 50 g of waste acid B. Thus, the mass fraction of phosphate acid and sulphuric acid was calculated to be 33.8% and 24.6%, respectively, in 50 g of waste acid B.

9.3.2.2 Calculation of Metal Element Content in Waste Acid

The total mass of each metal element within waste acid can be calculated based on the sediment and filtered solution, where the former is equal to the weight percentage of the metal element times the total sediment mass (based on Table 9.2), and the latter is equal to the mass fraction of the metal element times the total solution mass (based on Table 9.3). It is assumed that there are no impurities in the added calcium oxide. The calculated results for the selected six metal elements are summarised in Table 9.4, and these results were used as the base (the total metal elements within geopolymer) for calculating the solidification efficiency of heavy metals.

TABLE 9.4
Mass of the Selected Six Metal Elements in 50 g of Waste Acids A and B

	Element	In Sediment (g)	In Filtered Solution (g)	Total (g)
Waste acid A	Ag	0.0125	0.0000	0.0125
	Fe	0.0284	0.0002	0.0286
	Mn	0.0015	0.0002	0.0017
	Ni	0.0022	0.0001	0.0023
	Cu	0.0029	0.0012	0.0041
	Mo	0.3128	0.0006	0.3134
Waste acid B	Ag	0.0264	0.0000	0.0264
	Fe	0.0514	0.0002	0.0516
	Mn	0.0039	0.0004	0.0043
	Ni	0.0032	0.0004	0.0037
	Cu	0.0040	0.0025	0.0065
	Mo	0.0000	0.0001	0.0001

Source: Reprinted with permission from ref. [20]. Copyright 2024, Elsevier.

9.3.2.3 Heavy Metal Solidification Efficiency

Table 9.5 shows the selected six metal elements leached from the geopolymer. The heavy metal solidification efficiency of geopolymer can then be calculated using Equation (9.1), and the results are summarised in Table 9.6. It can be seen that geopolymers activated with both waste acid A and acid B show excellent heavy metal solidification efficiency. The majority of the heavy metal elements can be solidified successfully with over 90% solidification efficiency. The heavy metal concentrations are significantly lower than the standard's recommendation (HJ781-2016), indicating that the geopolymers are appropriate for hazardous material stabilisation and that the leaching of the heavy metals into the environment can be well controlled [26].

$$\text{Solidification efficiency} = \left(1 - \frac{\text{Mass leached from geopolymer}}{\text{Total mass in waste acid}}\right) \times 100\% \quad (9.1)$$

TABLE 9.5
Metal Elements Leached from Geopolymer via ICP (50 g of Geopolymer in 500 ml Water)

	μg/ml					
Element	**Ag**	**Cu**	**Fe**	**Mn**	**Mo**	**Ni**
WPAG-A	0.0001	1.297	2.321	0.175	1.338	0.099
WPAG-B	0.0001	3.100	1.348	0.499	0.001	0.151

Source: Reprinted with permission from ref. [20]. Copyright 2024, Elsevier.

TABLE 9.6
Heavy Metal Solidification Efficiency

Element	Ag	Fe	Mn	Ni	Cu	Mo
WPAG-A	100.00%	95.94%	94.85%	97.82%	84.32%	99.79%
WPAG-B	100.00%	98.69%	94.20%	97.94%	75.97%	99.65%

Source: Reprinted with permission from ref. [20]. Copyright 2024, Elsevier.

Traditionally, limestone is added to waste acid. The sediment is buried in a landfill, which cannot be used because of heavy metal contamination. The current technology allows the fixation of heavy metal elements in bricks, which can be used in pavement. At the end of service life, the heavy metal elements are diluted in the soil, which is less harmful to the land. Thus, the environmental impact of the geopolymer activated with waste acid is minimal. The heavy metal elements are distributed in the geopolymer in the form of Ag, Fe, Cu, Mo, and Ni, and the resultant material can be used as a sustainable alternative to conventional building materials and fire-resistant materials.

9.3.3 Mechanical Properties of Waste Acid-Activated Geopolymer

Table 9.7 summarises the tensile and flexural strength of WPAG-A and WPAG-B, and the compressive and flexural stress-strain curves of WPAG-A and WPAG-B are shown in Figure 9.6a and b, respectively. Results show that WPAG-A exhibits superior mechanical performance compared with WPAG-B, and the compressive strength of WPAG-A can reach 32.2 MPa. However, both samples show a lower flexural strength, which could be attributed to severe cracks forming during the ageing stage (Figure 9.3b).

The Headspace Gas Chromatography test confirms the presence of acetic acid in waste acid A. Previous research suggested that the CH_3COO^- and H^+ produced by ionisation of acetic acid (CH_3COOH) will react (acid-base reaction) with metal oxides in Metakaolin, generating $(CH_3COOH)_2Ca$, $Al(CH_3COOH)_3$, and $Fe(CH_3COOH)_3$. The acid-base reaction causes an uneven surface of the metakaolin particles, reducing the particle diameter; thus, the particles are well dispersed. As a result, the contact area between solid particles and solution increases, so as the hydration rate [27].

TABLE 9.7
Mechanical Properties of Geopolymer with Waste Acid A and Waste Acid B

Sample	Compressive Strength (MPa)	Young's Modulus (MPa)	Flexural Strength (MPa)	Young's Modulus (MPa)
WPAG-A	32.2 ± 2.2	1141 ± 70	2.7 ± 0.4	981 ± 250
WPAG-B	21.7 ± 2.0	743 ± 119	1.7 ± 0.3	413 ± 237

Source: Reprinted with permission from ref. [20]. Copyright 2024, Elsevier.

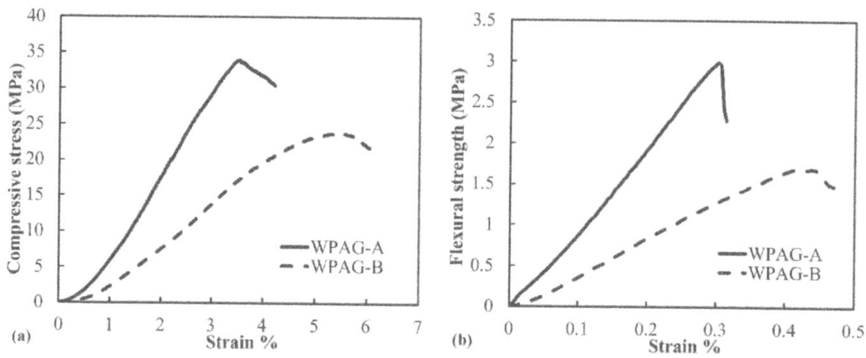

FIGURE 9.6 (a) Compressive and (b) flexural stress-strain curve of geopolymer activated with waste acid A and waste acid B. (Reprinted with permission from ref. [20]. Copyright 2024, Elsevier.)

A considerable amount of hydration products, such as C-S-H gel and CH, are produced. This causes a dense structure with less porosity. Besides, Ca^{2+}, Al^{3+}, and Fe^{3+} can be decomposed from $(CH_3COOH)_2Ca$, $Al(CH_3COOH)_3$, and $Fe(CH_3COOH)_3$, which can provide the necessary ions for the hydration reaction and improve the degree of hydration reaction [27].

Additionally, high concentrations of sulphuric acid were detected via the ICP test in waste acid B. It was suggested that a high concentration of sulphuric acid significantly reduced the pH value of the internal environment of the geopolymer slurry [27]. A lower pH value of the solution leads to lower monomer concentration, which could inhibit the polymerisation reaction [28]. Therefore, the compressive and flexural strength of WPAG-A is superior to that of WPAG-B.

The compressive and flexural strain-strain curves of WPAG-A and WPAG-B are plotted in Figure 9.6a and b, respectively. In both plots, the ascent and descent stages are clearly shown. It is also worth mentioning that the stress-strain curve slope of waste acid A is steeper compared to that of sample B, meaning that the modulus of elasticity increased. Thus, the ductility of the material is improved.

9.3.4 Thermal Gravimetric Analysis

Figure 9.7 presents the TG curve of WPAG-A and WPAG-B in the temperature range from 30°C to 1500°C. An extensive weight loss was observed in the temperature range from 30°C to 200°C, which represents the removal of absorbed water and hydroxyl groups [22]. The weight loss at 200°C of the WPAG-B is approximately 3% higher than WPAG-A, and this may be due to a higher water evaporation rate caused by a loose and open porous structure, as presented in Figure 9.3d. In the temperature range between 400°C and 1500°C, the mass loss of WPAG-A is almost unchanged, indicating the geopolymer activated with acid A exhibits excellent thermal stability and is thermally stable at a temperature of 1500°C. In contrast, an additional weight loss of 6% for WPAG-B was shown between 600°C and 800°C due to dehydroxylation, indicating moderate thermal stability performance [29].

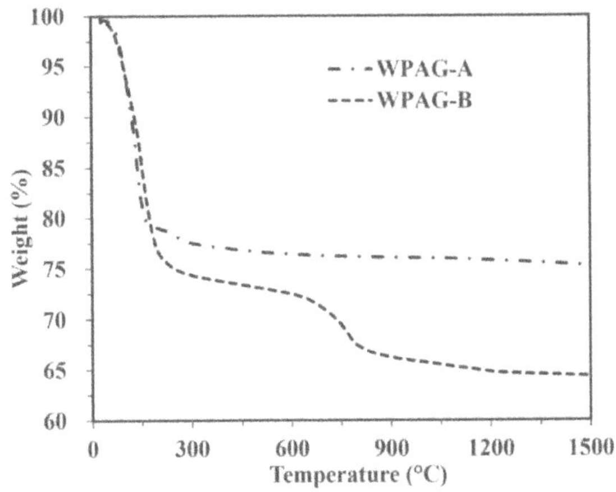

FIGURE 9.7 TGA curve of WPAG-A and WPAG-B. (Reprinted with permission from ref. [20]. Copyright 2024, Elsevier.)

9.3.5 Economic and Environmental Analysis

Table 9.8 summarises the CO_2 emissions, energy consumption, and cost of the common raw materials during production, and the transportation process is not considered. Cement has significantly higher CO_2 emissions and energy consumption compared to fly ash (FA) and MK. Since FA is a by-product during energy generation using coal, all three factors are the lowest. In addition, alkaline activators, such as NaOH and Na_2SiO_3, exhibit higher CO_2 emissions and energy consumption than phosphate acid, whereas the price of phosphate acid is significantly higher [19]. It

TABLE 9.8
CO_2 Emissions, Energy Consumption, and Cost of Raw Materials (Transportation Process Is Not Considered) [19, 30–34]

Constituent	CO_2 Emissions (kg/t)	Energy Consumption (GJ/t)	Cost ($/t)[a]
FA	12	0.173	45
Cement	730	4.5	150
MK	92	2.5	142.8[c]
PA solution	314.4	1.17	812.76
NaOH	1915	2.8	147.8
Na_2SiO_3	1222	5.371	264.4
Waste acid (WPA)[b]	0	0	–1428.6

Source: Reprinted with permission from ref. [20]. Copyright 2024, Elsevier.

[a] Market price may fluctuate.

[b] The CO_2 emission and the energy consumption are assumed as 0 since there is no additional treatment process after collecting the waste acid; the cost was quoted from the waste acid supplier.

[c] The price of Metakaolin was averaged based on local suppliers.

is worth mentioning that the CO_2 emission and energy consumption for producing waste acid is assumed to be 0 since there is no additional treatment process after collecting the waste acid. According to the supplier, there is a cost to recycle the waste acid from the enterprise that generates waste acid. Thus, the cost for waste acid is negative for distinguishing commercial phosphate acid and alkaline activators.

To compare the CO_2 emissions, energy consumption, and cost between various types of geopolymers and cement, the intensity (per m^3 sample per MPa) was used, and the results are summarised in Table 9.9 and Figure 9.8. It is well indicated that

TABLE 9.9
Typical Mixture Proportions of WPAG (This Work), PAFG (Phosphate-Activated Fly Ash), AAFG (Alkaline-Activated Fly Ash), PAMK (Phosphate-Activated Metakaolin), AAMK (Alkaline-Activated Metakaolin), and Portland Cement, as Well as Their CO_2 Emissions, Energy Consumption, and Cost

	WPAG-A[a]	WPAG-B[a]		AAFG [35]	AAFG [36]			
Type	Waste Acid A	Waste Acid B	PAFG [19]	NaOH	NaOH + Na_2SiO_3	PAMK [37]	AAMK [34]	PC Paste [38]
FA (kg/m^3)	–	–	1538.46	2127.66	1910.83	–		–
Cement (kg/m^3)	–	–	–	–	–	–		1333.33
MK (kg/m^3)	1262.37	1262.37	–	–	–	1302.08	858.8	–
PA (kg/m^3)	126.24	378.71	172.09	–	–	651.56	–	–
WPA (kg/m^3)	1136.14	883.66	–	–	–	–	–	–
NaOH (kg/m^3)	–	–	–	223.40	43.35	–	159.17	–
Na_2SiO_3 (kg/m^3)	–	–	–	–	183.44	–	945.07	–
Water (kg/m^3)	252.47	378.71	366.37	414.89	537.54	618.41	–	800
Compressive strength (MPa)	32.20	21.70	13.23	17.3	7.35	17.90	51.35	12.3
Carbon emission (kg/m^3)	187.20	269.87	230.91	453.35	666.36	340.75	705.04	973.33
Energy consumption (GJ/m^3)	3.47	3.78	1.37	1	2.91	4.23	8.54	6
Cost ($/m^3)	−1319.57	−745.19	255.38	128.76	213.67	760.76	401.27	200
Carbon intensity (kg/m^3/MPa)	5.81	12.44	17.45	26.21	90.66	19.04	13.73	79.13
Energy intensity (MJ/m^3/MPa)	107.89	174.32	103.34	57.69	396	236.12	166.31	487.8
Cost intensity ($/m^3/MPa)	−40.98	−34.34	19.3	7.44	29.07	42.50	7.81	16.26

Source: Reprinted with permission from ref. [20]. Copyright 2024, Elsevier.

[a] According to He et al. [19], the power consumption of an oven at 60°C for 24 h is 40 kWh. The CO_2 emissions, energy consumption, and costs per kWh of electricity are 0.62 kg, 3.6 MJ, and $0.09, respectively.

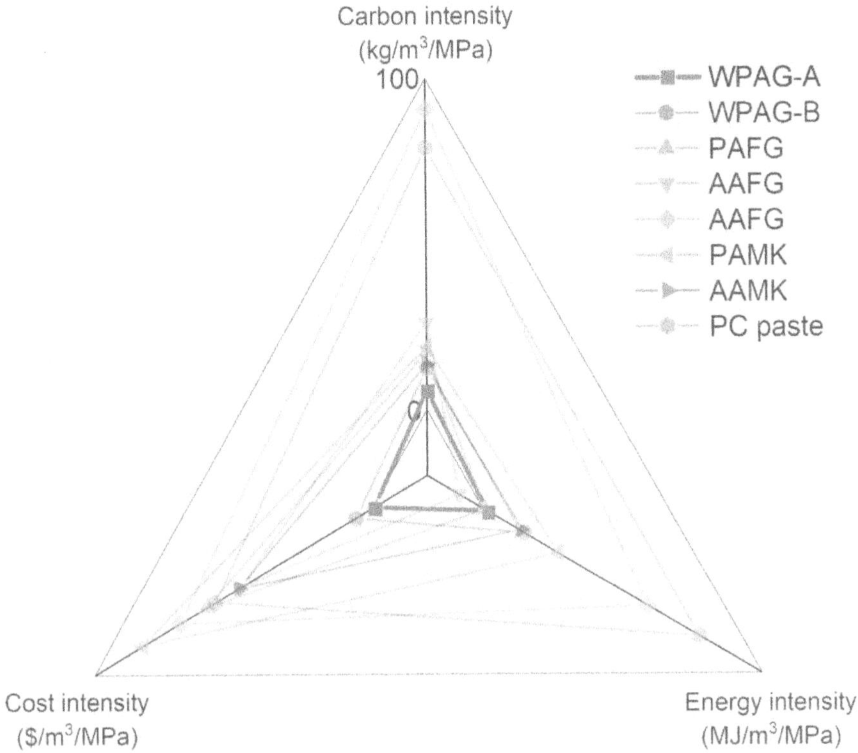

FIGURE 9.8 CO_2 emissions, energy consumption, and cost intensity comparison of various geopolymers. (Reprinted with permission from ref. [20]. Copyright 2024, Elsevier.)

the constituents of the geopolymer have a significant influence on CO_2 emissions, energy consumption, and cost. Apart from cement, the geopolymer activated with NaOH and Na_2SiO_3 has the highest carbon intensity and energy consumption intensity due to the significantly high carbon emission and energy consumption of the NaOH and Na_2SiO_3 during the production process among all listed samples. The FA-based geopolymer activated with NaOH exhibits the lowest energy consumption intensity due to the use of fly ash and a higher amount of water. It can be seen that the carbon intensity of the sample prepared in this work is substantially lower than that shown in the literature due to the use of waste acid as the activator. The carbon intensity was reduced by 92.7% and 57.7% compared to cement and AAMK, respectively (the lowest reported value), as listed in Table 9.9. In addition, the cost intensity is negative due to the income of waste acid treatment, meaning that additional value benefit was added to the product. This paves a promising way for the low-carbon transformation of ceramic enterprises and industrial waste treatment enterprises, obtaining greater economic and social benefits for related enterprises, driving further breakthroughs in high-tech industries, and promoting the sustainable development of the local economy. Nevertheless, the energy intensity is higher than the AAFG (alkaline-activated fly ash) (the lowest reported value) as the calcining of kaolin requires high energy consumption. As a result, the energy intensity is 87%

higher than AAFG, although it is already 77.9% lower than the PC paste. Therefore, in the future, an investigation will be carried out by replacing Metakaolin with fly ash to maximise the environmental and economic benefits.

9.4 CONCLUSIONS

This chapter investigated the feasibility of using industrial waste acid to prepare geopolymers with environmental and economic benefits. Two types of waste acid were examined to discuss the effect of waste acid content on the microstructure, mechanical performance, thermal stability, and heavy metal solidification efficiency. The core issue solved in this investigation is the harmless treatment of industrial waste acid and solidification efficiency of heavy metals, so as to reduce the pollution of heavy metals on the environment and the urgent demand for low carbon emissions in the construction and ceramic industries. Several conclusions can be drawn from the study:

1. The geopolymer activated with waste acid containing phosphate and acetic acid (waste acid A) exhibited a consolidated structure, high compressive strength (32.2 MPa), and excellent thermal stability. This may be because the ionisation of acetic acid promotes the dispersion of solid particles. As a result, the contact area between solid particles and solution increases, so as the hydration rate.
2. The heavy metal solidification test showed that the metal elements (Ag, Fe, Cu, Mo, and Ni) can be successfully solidified with over 95% solidification efficiency for both samples, indicating that the geopolymers are appropriate for hazardous material stabilisation, and the leaching of the heavy metals to the environment can be well controlled.
3. The potential CO_2 emissions, energy consumption, and cost comparison between various geopolymers showed that the carbon intensity of the samples prepared in this study reduced by 92.7% and 57.7% compared to cement and AAMK, respectively (the lowest reported value listed). The energy intensity was 77.9% lower than the PC paste. What is more important is that the cost intensity was negative due to the income of waste acid treatment, meaning that additional value benefit is added to the product.

These findings pave a promising way for the low-carbon transformation of ceramic enterprises and industrial waste treatment enterprises, drive further breakthroughs in high-tech industries, and promote the sustainable development of the local economy. Meanwhile, further investigation can be conducted:

1. Notably, the CO_2 emissions, energy consumption, and cost of MK are significantly higher than the industrial waste, such as fly ash; therefore, it would be more beneficial to develop a joint treatment technology for industrial waste acid and solid waste.
2. The flexural strength of the samples prepared is relatively low due to severe cracks exhibited during the ageing stage; thus, the fabrication process could be optimised by introducing the step moisture ageing.

3. The reaction mechanisms between waste acids, such as phosphate acid, acetic acid, sulphuric acid, and aluminosilicate solids (especially fly ash and other industrial solid waste) are still unclear. Thus, further research is needed, as it plays a key role in the industrialisation of the combined treatment of industrial waste acid and industrial solid waste in the later stage.

ACKNOWLEDGEMENT

The financial support of Tsinghua-Foshan Advanced Manufacturing Research Institute is acknowledged. The authors are grateful for the support of colleagues at the Future Materials & Design Research Center, The Future Lab, Tsinghua University. The assistance of the technical staff at the School of Materials Science and Engineering Analysis Center, Tsinghua University is also acknowledged.

REFERENCES

1. Y. Zhang *et al.*, "Removal of chloride from waste acid using Bi_2O_3: Thermodynamics and dechlorination behavior," *Journal of Water Process Engineering*, vol. 49, p. 103048, 2022.
2. Q. Liu, Y. Liu, G. Zhao, and T. Yang, "Research progress on solidification of heavy metal ions by geopolymer," *Applied Chemical Industry*, vol. 51, no. 4, 2022.
3. Z. Yuan, L. Ze, and W. Dongmin, "Research progress of the stabilization of heavy metals using fly ash based geopolymer" (in Chinese), *Bulletin of the Chinese Ceramic Society*, vol. 35, no. 6, 2016.
4. S. Wang *et al.*, "Application of geopolymers for treatment of industrial solid waste containing heavy metals: State-of-the-art review," *Journal of Cleaner Production*, vol. 390, 2023.
5. K. Wang and H. Le, "The development of cement-based, intumescent and geopolymer fire-retardation coatings for metal structures: A review," *Coatings*, vol. 13, no. 3, 2023.
6. M. Lahoti, K. H. Tan, and E.-H. Yang, "A critical review of geopolymer properties for structural fire-resistance applications," *Construction and Building Materials*, vol. 221, pp. 514–526, 2019.
7. P. Arokiasamy *et al.*, "Diverse material based geopolymer towards heavy metals removal: A review," *Journal of Materials Research and Technology*, vol. 22, pp. 126–156, 2023.
8. R. M. Novais, L. H. Buruberri, M. P. Seabra, and J. A. Labrincha, "Novel porous fly-ash containing geopolymer monoliths for lead adsorption from wastewaters," *Journal of Hazardous Materials*, vol. 318, pp. 631–640, 2016.
9. N. Grba, A. Baldermann, and M. Dietzel, "Novel green technology for wastewater treatment: Geo-material/geopolymer applications for heavy metal removal from aquatic media," *International Journal of Sediment Research*, vol. 38, no. 1, pp. 33–48, 2023.
10. T. Lan *et al.*, "Mixed precursor geopolymer synthesis for removal of Pb(II) and Cd(II)," *Materials Letters*, vol. 274, p. 127977, 2020.
11. S. Yan *et al.*, "A green and low-cost hollow gangue microsphere/geopolymer adsorbent for the effective removal of heavy metals from wastewaters," *Journal of Environmental Management*, vol. 246, pp. 174–183, 2019.
12. B. Kim, J. Kang, Y. Shin, T.-m. Yeo, and W. Um, "Immobilization mechanism of radioactive borate waste in phosphate-based geopolymer waste forms," *Cement and Concrete Research*, vol. 161, 2022.

13. S. Pu, Z. Zhu, W. Song, H. Wang, W. Huo, and J. Zhang, "A novel acidic phosphoric-based geopolymer binder for lead solidification/stabilization," *Journal of Hazardous Materials*, vol. 415, p. 125659, 2021.
14. H. Guo, B. Zhang, L. Deng, P. Yuan, M. Li, and Q. Wang, "Preparation of high-performance silico-aluminophosphate geopolymers using fly ash and metakaolin as raw materials," *Applied Clay Science*, vol. 204, 2021.
15. H. Lin, H. Liu, Y. Li, and X. Kong, "Thermal stability, pore structure and moisture adsorption property of phosphate acid-activated metakaolin geopolymer," *Materials Letters*, vol. 301, 2021.
16. C. Nobouassia Bewa, H. K. Tchakouté, D. Fotio, C. H. Rüscher, E. Kamseu, and C. Leonelli, "Water resistance and thermal behavior of metakaolin-phosphate-based geopolymer cements," *Journal of Asian Ceramic Societies*, vol. 6, no. 3, pp. 271–283, 2018.
17. Y.-S. Wang, Y. Alrefaei, and J.-G. Dai, "Silico-aluminophosphate and alkali-aluminosilicate geopolymers: A comparative review," *Frontiers in Materials*, vol. 6, 2019.
18. Y. Liu *et al.*, "Novel porous phosphoric acid-based geopolymer foams for adsorption of Pb(II), Cd(II) and Ni(II) mixtures: Behavior and mechanism," *Ceramics International*, vol. 49, no. 4, pp. 7030–7039, 2023.
19. M. He, Z. Yang, N. Li, X. Zhu, B. Fu, and Z. Ou, "Strength, microstructure, CO_2 emission and economic analyses of low concentration phosphoric acid-activated fly ash geopolymer," *Construction and Building Materials*, vol. 374, 2023.
20. K. Wang, Q. Wei, H. Chen, and H. Le, "Solidification of heavy metal elements in waste phosphate acid activated metakaolin geopolymer with environmental and economic benefit," *Ceramics International*, vol. 50, no. 13, Part A, pp. 22874–22883, 2024.
21. Q. R. Institute, "Market prospect forecast and investment strategy planning analysis report of waste acid recycling industry in China," 2020. Available: https://bg.qianzhan.com/trends/detail/506/200713-a7801351.html.
22. L. Gao, Y. Zheng, Y. Tang, J. Yu, X. Yu, and B. Liu, "Effect of phosphoric acid content on the microstructure and compressive strength of phosphoric acid-based metakaolin geopolymers," *Heliyon*, vol. 6, no. 4, p. e03853, Apr. 2020.
23. S. Petlitckaia and A. Poulesquen, "Design of lightweight metakaolin based geopolymer foamed with hydrogen peroxide," *Ceramics International*, vol. 45, no. 1, pp. 1322–1330, 2019.
24. H. K. Tchakouté, C. H. Rüscher, E. Kamseu, F. Andreola, and C. Leonelli, "Influence of the molar concentration of phosphoric acid solution on the properties of metakaolin-phosphate-based geopolymer cements," *Applied Clay Science*, vol. 147, pp. 184–194, 2017.
25. C.-m. Guo, K.-t. Wang, M.-y. Liu, X.-h. Li, and X.-m. Cui, "Preparation and characterization of acid-based geopolymer using metakaolin and disused polishing liquid," *Ceramics International*, vol. 42, no. 7, pp. 9287–9291, 2016.
26. H. Majdoubi *et al.*, "Valorization of phosphogypsum waste through acid geopolymer technology: Synthesis, characterization, and environmental assessment," *Construction and Building Materials*, vol. 371, 2023.
27. Y. Guihong, "Mechanical and Shrinkage Properties of Acid Activated Cement-Based Materials," Master thesis, China University of Mining and Technology, 2018.
28. D. Khale and R. Chaudhary, "Mechanism of geopolymerization and factors influencing its development: A review," *Journal of Materials Science*, vol. 42, no. 3, pp. 729–746, 2007.
29. L. Le-ping, C. Xue-min, Q. Shu-heng, Y. Jun-li, and Z. Lin, "Preparation of phosphoric acid-based porous geopolymers," *Applied Clay Science*, vol. 50, no. 4, pp. 600–603, 2010.
30. R. Gilmour, *Phosphoric acid: purification, uses, technology, and economics*. CRC Press, 2013.

31. L. K. Turner and F. G. Collins, "Carbon dioxide equivalent (CO_2-e) emissions: A comparison between geopolymer and OPC cement concrete," *Construction and Building Materials*, vol. 43, pp. 125–130, 2013.
32. G. F. Huseien, A. R. M. Sam, K. W. Shah, J. Mirza, and M. M. Tahir, "Evaluation of alkali-activated mortars containing high volume waste ceramic powder and fly ash replacing GBFS," *Construction and Building Materials*, vol. 210, pp. 78–92, 2019.
33. I. Faridmehr, M. L. Nehdi, M. Nikoo, G. F. Huseien, and T. Ozbakkaloglu, "Life-cycle assessment of alkali-activated materials incorporating industrial byproducts," *Materials*, vol. 14, no. 9, p. 2401, 2021.
34. P. Perez-Cortes and J. I. Escalante-Garcia, "Alkali activated metakaolin with high limestone contents – Statistical modeling of strength and environmental and cost analyses," *Cement and Concrete Composites*, vol. 106, p. 103450, 2020.
35. A. Palomo, M. W. Grutzeck, and M. T. Blanco, "Alkali-activated fly ashes: A cement for the future," *Cement and Concrete Research*, vol. 29, no. 8, pp. 1323–1329, 1999.
36. S. Sasui, G. Kim, J. Nam, T. Koyama, and S. Chansomsak, "Strength and microstructure of class-C fly ash and GGBS blend geopolymer activated in NaOH & NaOH + Na_2SiO_3," *Materials*, vol. 13, no. 1, p. 59, 2020.
37. X. Xie, T. Liu, J. Zhang, Y. Zheng, and J. Gao, "Effect of acid-activator characteristics on the early hydration behavior and properties of metakaolin-based geopolymer," *Journal of Building Engineering*, vol. 72, 2023.
38. B. Fang, T. Xu, and S. Shuang, "Laboratory study on cement slurry formulation and its strength mechanism for semi-flexible pavement," *Journal of Testing and Evaluation*, vol. 44, pp. 907–913, 2016.

10 Emerging Applications of Geopolymers in Water Treatment

Mohammad Rauf, Kaibao Wang, and Huirong Le

10.1 INTRODUCTION

Davidovits proposed the name of geopolymers in the 1970s, and they have become an important class of materials because they provide a sustainable substitute for conventional materials like cement and some ceramics [1, 2]. These compounds consist of aluminosilicates (such as fly ash, slag, or metakaolin) combined with alkaline solutions (like sodium hydroxide) or phosphoric acid [3]. In addition to being more environmentally friendly than traditional materials, geopolymers are notable for their superior mechanical strength, exceptional chemical resistance, and outstanding thermal stability [4, 5]. Their possibility to be generated with less energy makes them desirable for high-performance applications like filtration, thermal insulation, and adsorption as well as for use in building materials. Because of their high porosity (can be as high as 50%), porous geopolymers have drawn interest recently because of their potential applications in a number of industries, including gas separation, wastewater treatment, and catalyst scaffolds [6].

The study and publication of articles on geopolymer-based membranes appear to have increased as research has progressed, especially in the past 10 years [7, 8]. The difficulties with conventional ceramic membranes, which require costly and energy-intensive procedures like sintering and calcination, cause this extension. In many industrial processes, ceramic membranes are chosen over their polymeric counterparts due to their resistance to strong chemicals and high temperatures [9]. However, they are expensive to produce and require raw elements like titania and zirconia.

Figure 10.1 displays the research trends for conventional ceramic membranes and geopolymer membranes between 2017 and 2024. According to the data, interest in and publications about geopolymer membranes have rapidly increased because of their cost-effectiveness and sustainability. In contrast, research on conventional ceramic membranes has stayed mostly unchanged [10, 11]. In contrast, the geopolymerisation technique makes it possible to create high-performing membranes without these limitations. Because using industrial waste materials like fly ash lowers production costs and reduces environmental impact, geopolymer membranes are a desirable substitute [12]. This shift has led to a steady rise in research publications since 2011, with the focus on

DOI: 10.1201/9781003535683-10

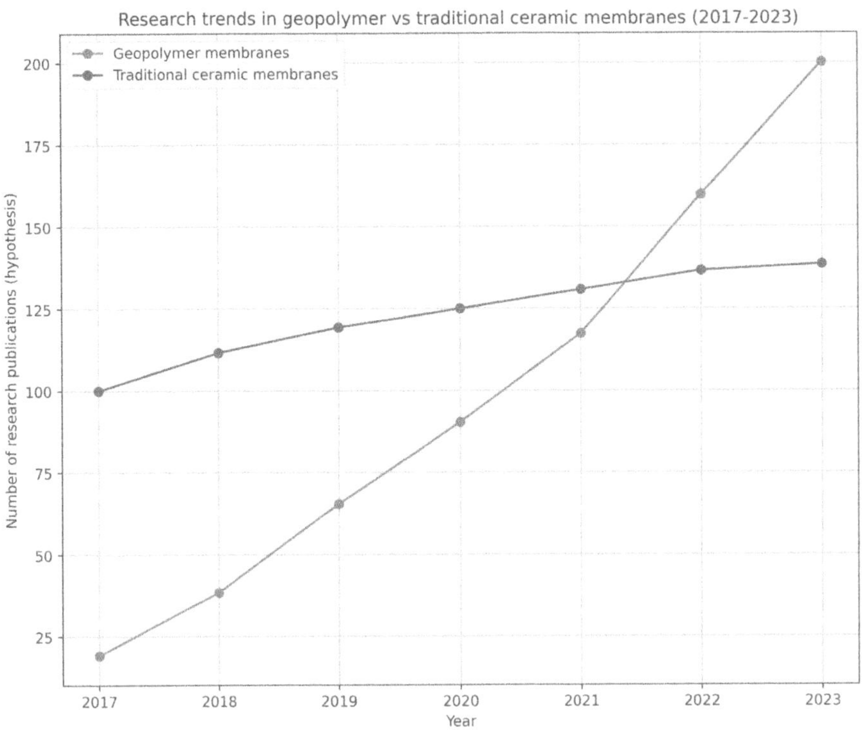

FIGURE 10.1 Research trends in geopolymer vs traditional ceramic membranes.

using geopolymer membranes growing more decisive year by year, as shown in Figure 10.1.

The potential of geopolymer membranes offers a more economical and environmentally friendly way to meet the high needs of industries that depend on water purification, gas separation, or filtration. Traditional membranes, especially ceramic ones, have long been preferred when polymeric membranes fail in high-pressure, high-temperature, and chemically harsh conditions. Nevertheless, producing these ceramic membranes continues to be expensive and energy-intensive. By providing a means of avoiding the sintering process while preserving strength and stability, geopolymer membranes offer a much-needed substitute. Researchers have demonstrated that the pore size and shape of these membranes can be altered, which is essential for filtration. They have shown that in some applications, including wastewater treatment, where eliminating impurities is crucial, geopolymer membranes can perform on par with or even better than conventional ceramic membranes.

Despite the progress, there are still some challenges to be addressed. One key issue is scaling up the production of geopolymer membranes to industrial levels while maintaining consistent pore size and mechanical properties across large areas. Researchers are still fine-tuning the process to optimise the membranes' performance, especially regarding long-term durability and efficiency in large-scale applications. Nevertheless, the potential of geopolymer membranes is undeniable. They provide a sustainable path forward by turning industrial waste, such as fly ash, into valuable materials that can be used to tackle global environmental issues, from reducing water pollution to minimising energy consumption [13].

The study of fly ash-based geopolymer membranes has gained increasing attention since 2022. One advantage of fly ash, a plentiful by-product of burning coal, is that it may be used to create inexpensive, high-performing membranes while also reducing the environmental problems associated with its disposal [14]. By modifying their composition and pore structure, fly ash-based membranes can be made to fit particular uses. As companies look for more effective and sustainable ways to meet their material needs, the number of publications on geopolymer membranes is clearly increasing. Geopolymer membranes may play a key role in filtration technologies in the future, with more study and improvement, assisting sectors in their transition to more ecologically friendly, circular operations.

10.2 RESEARCH PROGRESS

10.2.1 Raw Material

Table 10.1 summarises the raw materials suitable for the making of traditional and geopolymer membranes for water filtration application. Traditional sintered ceramics and geopolymer membranes are made from various raw materials, each adding unique qualities that improve filtration efficiency. According to Goswami et al. [15], fly ash offers an affordable source of aluminosilicate for ceramic membranes used in wastewater treatment applications. Mortar et al. [11] assert that blast furnace slag

TABLE 10.1
List of Raw Materials Suitable for the Making of Traditional and Geopolymer Membranes for Water Filtration Application

Raw Material	References	Raw Material	References
Fly Ash	[15]	Metakaolin	[11]
Blast furnace slag	[11]	Red mud	[18]
Waste glass	[11]	Kaolin	[11]
Alumina (Al_2O_3)	[16]	Titanium dioxide (TiO_2)	[17]
Calcium carbonate	[17]	Polyethersulfone (PES)	[17]
Polyvinylidene fluoride (PVDF)	[17]	Cellulose acetate	[17]
Carbon nanotube	[15]	Zeolites	[16]
Graphene oxide	[17]	Silica fume	[18]
Sodium silicate	[16]	Sodium hydroxide (NaOH)	[11]
Potassium hydroxide (KOH)	[18]	Aluminum powders	[16]
Hydrogen peroxide (H_2O_2)	[18]	Zirconia (ZrO_2)	[11]
Sodium dodecyl sulphate (SDS)	[17]	Sodium lauryl sulphate (SLS)	[17]

and metakaolin are essential for improving the mechanical strength and thermal resistance of geopolymer ceramics. Red mud helps produce sustainable geopolymers as an alternate aluminosilicate precursor [18]. In ceramic geopolymer applications, waste glass and kaolin are also common materials; kaolin enhances chemical resistance [11]. According to Sadiq et al. [16] and Cheng et al. [17], traditional inorganic materials with excellent chemical resistance and structural stability, like alumina (AlO_3) and titanium dioxide (TiO_2), help improve membrane durability and environmental cleanup.

According to Cheng et al. [17], polymeric materials such as polyethersulfone (PES) and polyvinylidene fluoride (PVDF) are vital for improving membrane performance and flexibility in water treatment. Because of its outstanding permeability and biodegradability, cellulose acetate is also used as a functional base in membrane applications. Because of their large surface area and exceptional adsorption capabilities, advanced materials such as graphene oxide [17], zeolites [16], and carbon nanotubes [15] provide significant gains in filtration efficiency. By controlling porosity, silica fume helps improve the structural integrity of geopolymer membranes [18].

The geopolymerisation process depends on alkaline activators like sodium silicate and sodium hydroxide (NaOH), which maximise the chemical linkages in the membrane matrix [16, 11]. When used as an activator, potassium hydroxide (KOH) raises the reactivity of high-performance geopolymer membranes [18]. Furthermore, hydrogen peroxide (H_2O_2) and aluminium powders serve as pore-forming agents, generating regulated porosity for applications involving selective filtering [16, 18]. Zirconia (ZrO_2) improves membrane toughness and offers resistance to wear under severe mechanical loads, according to Mortar et al. [11]. By strengthening surface contacts and stability during filtration, surface-active substances such as sodium lauryl sulphate (SLS) and sodium dodecyl sulphate (SDS), as noted by Cheng et al. [17], further enhance membrane performance.

10.2.2 Processing Techniques

Table 10.2 lists the typical fabrication methods of geopolymer membranes. The advantages and disadvantages of each method are summarised. The direct foaming and dip-coating method is relatively straightforward, but the quality of products is generally hard to control. The extrusion method seems to be a suitable method for large-scale production, whereas intensive equipment and space are needed.

10.2.2.1 Uniaxial Dry Pressing Method

To build dense structures, dry powders combined with binders are pressed into a mould using unidirectional pressure in a simple procedure known as uniaxial dry pressing. To increase mechanical strength, the membranes are sintered and dried after pressing. According to Sadiq et al. [16], uniaxial pressing works well for simpler geometries with good mechanical stability, which makes it appropriate for common filtration applications that call for longevity (Figure 10.2).

TABLE 10.2
Membrane Fabrication Methods

Fabrication Method	Description	Common Membranes Produced	Advantages	Disadvantages	References
Direct Foaming	Introducing gas bubbles into a geopolymer slurry to create pores. Foaming agents like hydrogen peroxide or aluminum powder are often used	Fly ash-based geopolymers Composite geopolymer membranes	Simple method Pore size can be controlled with foaming agents	Pore distribution may be uneven Limited to larger pore sizes	[15, 18, 16]
Dip-Coating	A substrate (e.g., ceramic or metal mesh) is dipped into a geopolymer slurry, forming a thin layer on the surface. Multiple layers can be applied for thickness	Composite geopolymer membranes Zeolite-geopolymer membranes	Easy to control thickness Suitable for creating composite layers	May result in uneven coating Requires multiple layers for increased strength	[18, 15]
Extrusion Method	Used primarily for ceramic and geopolymer membranes, materials are extruded through a die to create hollow fibres or flat sheets	Fly ash-based ceramic membranes Zeolite membranes	Suitable for large-scale production Good control over shape and thickness	Equipment-intensive Limited to specific shapes (e.g., tubes, fibres)	[16, 19, 15]
Slip Casting	A slurry of ceramic or geopolymer material is poured into a mould where the liquid is drained off, leaving behind a cast structure	Ceramic membranes Geopolymer membranes	Allows complex shapes Good for creating thick membranes	Slow process Shrinkage during drying can affect the final size	[16, 19, 15]
Phase Inversion	After being cast onto a support and submerged in a coagulation bath, the polymer solution precipitates and forms a porous membrane	Polymeric membranes (PSF, PVDF) Composite polymer-inorganic membranes	Good control over pore size Widely used for ultrafiltration membranes	Requires precise control of phase separation Can lead to pore blockage if not carefully managed	[18, 15, 16]
Sintering	After being moulded, ceramic or geopolymer materials are heated to high temperatures (1200–1500°C) to solidify and form porous structures	Inorganic ceramic membranes Geopolymer membranes	Produces high-strength, durable membranes Can achieve small pore sizes	Energy-intensive Requires expensive raw materials like zirconia and alumina	[16, 20, 15]
Electrospinning	A polymer solution is electrostatically charged and spun into fine fibres collected on a surface to form a porous membrane	Nanofibre composite membranes Polymeric membranes	Produces nanofibre membranes Highly controllable pore sizes	Complex process Limited scalability	[17]

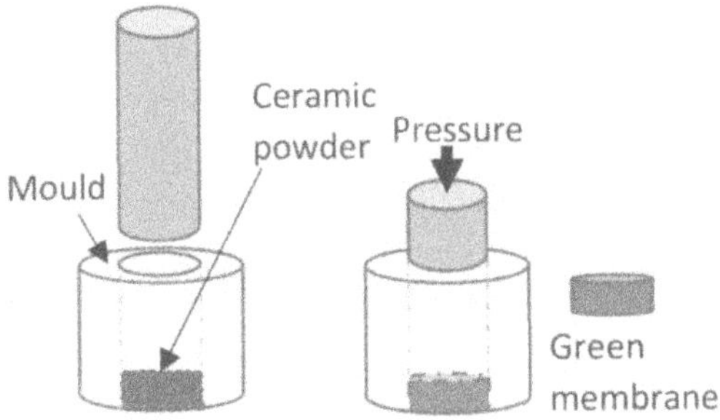

FIGURE 10.2 Illustration of a uniaxial dry pressing method [19].

10.2.2.2 Paste Casting Method

Paste casting produces membranes that are subsequently dried and sintered using moulds and a thick slurry composed of water and raw ingredients. Although careful drying is necessary to prevent cracks, this process is cost-effective and appropriate for building larger structures. In their review, Goswami et al. [15] emphasise the potential of paste casting to create thicker, more substantial membrane layers, which are perfect for structural applications in filtering systems (Figure 10.3).

10.2.2.3 Extrusion Method

Extrusion creates tubular or multi-channel membranes with high surface-to-volume ratios that are advantageous for filtration by pressing a thick paste through a die. The tubes are dried and sintered after extrusion. Because extrusion may produce complicated geometries with various flow channels, Mortar et al. [11] define it as an effective approach for high-flow applications. This technology is frequently employed in industrial water filtration installations (Figure 10.4).

10.2.2.4 Tape Casting

Tape casting creates thin, flat membrane layers that are sintered by spreading a slurry over a level surface, adjusting its thickness with a doctor's blade, and then drying it. Nurlina et al. [18] highlighted the accuracy of tape casting, which makes it appropriate for layered filtration systems that need consistency in thickness and porosity for high filtration selectivity (Figure 10.5).

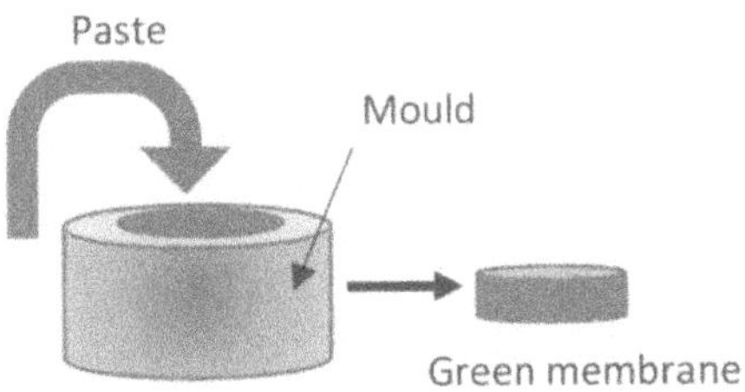

FIGURE 10.3 Illustration of the paste casting method [19].

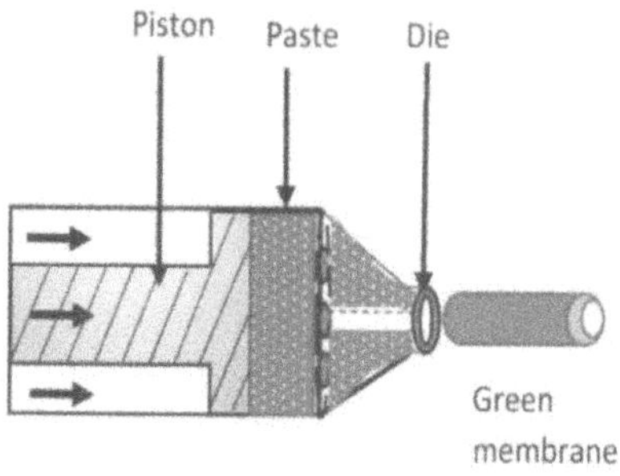

FIGURE 10.4 Illustration of the extrusion method [19].

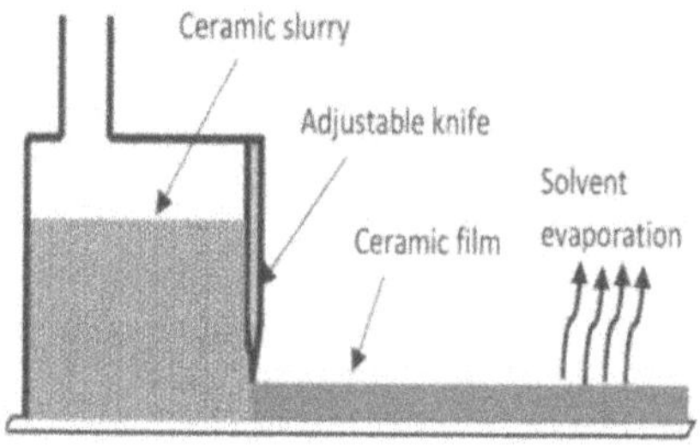

FIGURE 10.5 Illustration of tape method [19].

10.2.2.5 Freeze Casting

Directional porosity is created through a process known as freeze casting or ice templating. This method involves freezing a slurry, which leads to the formation of ice crystals that push particles closer together. Once the freezing process is complete, the ice sublimates, resulting in a porous structure that is then sintered. According to Sadiq et al. [16], freeze casting produces highly porous membranes with aligned, open channels, making them ideal for applications that require directional permeability (Figure 10.6).

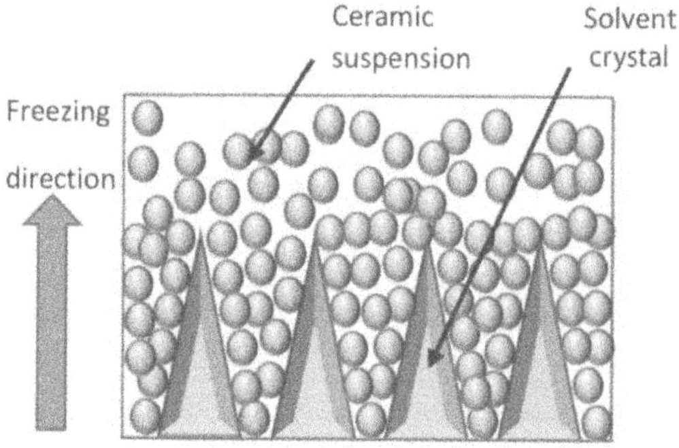

FIGURE 10.6 Illustration of freeze method [19].

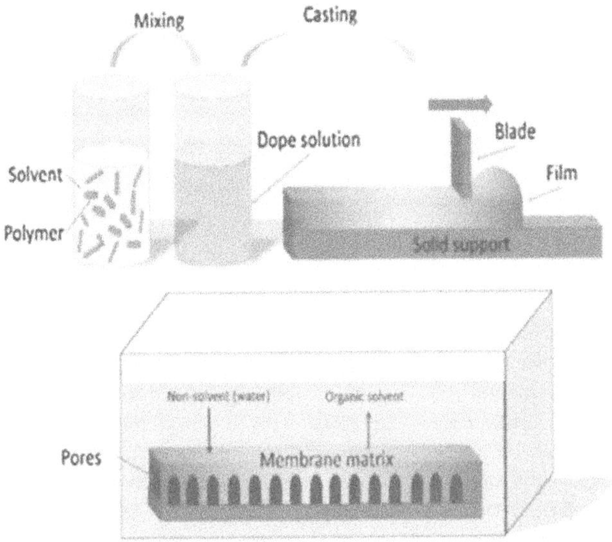

FIGURE 10.7 Illustration of phase inversion method [21].

10.2.2.6 Phase Inversion Casting

Phase inversion, commonly applied to polymeric membranes, creates an asymmetric structure with both dense and porous layers by casting a solution that undergoes phase separation in a coagulation bath. Cheng et al. [17] explain how manipulating the solution composition and bath conditions can effectively achieve selective permeability, especially for hollow-fibre membranes (Figure 10.7).

10.2.3 Applications of Geopolymer as Filter Materials

Geopolymers have been applied extensively in filtration technologies, including water and air purification, due to their high porosity and mechanical robustness. Because geopolymers can eliminate pollutants like heavy metals from industrial effluent, fly ash-based geopolymers are highly valued for water filtration. These membranes were ideal for industrial filtration applications because of their high porosity and mechanical strength, which included compressive values of up to 18.5 MPa. Metakaolin-based self-supported polymeric membranes for wastewater filtration were also investigated by Xu et al [23]. These membranes were successful in eliminating nanoparticles because they have pore diameters ranging from 20 to 100 nm. These membranes performed exceptionally well in water filtration jobs, which makes them useful for microfiltration and ultrafiltration applications.

Geopolymer membranes have also been utilised in air filtration, particularly for capturing particulate matter (PM) such as PM2.5 and PM10. A porous gradient geopolymer membrane developed by Wang et al. showed high removal efficiency for PM2.5 (98.7%) and PM10 (99.5%), demonstrating their effectiveness in industrial air filtration applications [24] .

TABLE 10.3
Types of Membranes

Type of Membrane	Material Used for Fabrication	Advantages	Disadvantages	References
Ceramic Membrane	Alumina (Al_2O_3), zirconia (ZrO_2), titania (TiO_2), silicon carbide (SIC)	High thermal and chemical stability Lower fouling rate Easy to clean	High cost Energy intensive synthesis	[16] [19] [15]
Polymeric Membrane	PVDF, PSF, PES, polyimides, polyamide, nanocomposites	Wide range of applications Low cost Easy to fabricate	Prone to fouling Non-biodegradable Environmental concerns	[22, 18] [16] [20, 15]
Geopolymer Membrane	Fly ash, metakaolin, red mud, aluminosilicates, Na_2SiO_3, NaOH	Low Cost Energy efficient synthesis Good thermal and chemical stability	Reproducibility issues Moderate fouling resistance	[18, 15] [20]
Composite Geopolymer Membrane	Geopolymer with fillers like zeolite, TiO_2, and fibres	Improved mechanical strength Better selectivity and resistance	Complex fabrication Costly additives	[18, 15]
Inorganic Membrane	Kaolinite, mullite, fly ash, silicon carbide (SIC)	High stability in harsh conditions Long durability	Expensive raw materials Requires high sintering temperature	[16] [20, 15]
Photocatalytic Geopolymer	Geopolymer with TiO_2, Cr_2O_3	Good for pollutant degradation Photocatalytic activity	Limited application More research needed	[18, 15]

It is also worth mentioning that the use of waste materials to treat wastewater provides a promising future application. Geopolymer-zeolite composite membranes have been successfully applied to remove heavy metals from wastewater, including chromium. The zeolite component provides a high surface area, enhancing the adsorption capabilities of the membrane, while the geopolymer matrix provides mechanical stability. This makes these composite materials excellent candidates for treating complex wastewater streams.

The performance of geopolymers in filtration applications is highly dependent on their microstructure, which includes porosity, pore size distribution, and mechanical strength. Tailoring these characteristics allows for optimisation in filtration efficiency. Frequently, foaming chemicals like hydrogen peroxide and starch are used to regulate the porosity of geopolymer membranes. For instance, a study by Naveed et al. [25] showed that overall porosity increased with increasing hydrogen peroxide concentrations, reaching 53.21% at the highest concentration. The trade-off between strength and filtering efficiency was highlighted by the fact that this increase in porosity was linked to a loss in compressive strength. Similarly, using starch as a foaming agent produced porosity levels of up to 51.09% with pore diameters ranging from 1.77 to 3.45 μm, which is perfect for applications in water filtering [26]. SDS, as a stabilising agent, plays a key role in regulating the porosity and stability of geopolymer foams by reducing surface tension and promoting pore uniformity. Korat and Ducman showed that adding SDS to alkali-activated fly ash geopolymers resulted in more homogeneous pore structures, which improved the mechanical properties of the final product [27].

Chitosan-modified geopolymer sub-microparticles were utilised in work by Song et al. to control the pore structure of geopolymer membranes, resulting in pore diameters as small as 58 nm. These membranes showed promise in water treatment by removing heavy metals and microplastics with high removal efficiency (89.37%) [28]. When evaluating 'geopolymers' appropriateness for filtering applications, their mechanical characteristics, in particular, their compressive strength, are quite important. Fly ash-based geopolymers, for example, showed compressive strengths of up to 18.5 MPa, which indicates that they are robust enough for high-pressure filtration applications. Nevertheless, depending on the application, a trade-off that must be controlled is that higher porosity frequently leads to reduced mechanical strength. By adding further layers, like a chitosan coating, geopolymer membranes can be made even stronger. By mixing chitosan with porous geopolymers, Zhang et al. created an inorganic-organic composite membrane that has a high water flux and enhanced pollutant removal efficiency [29]. Because of their exceptional chemical and thermal stability, geopolymers can withstand harsh conditions. For example, potassium-based geopolymers are perfect for high-temperature filtration applications since they can tolerate temperatures of up to 800°C. Geopolymers are also extremely resistant to basic and acidic conditions, which increases their suitability for use in challenging industrial situations.

One recent case study is based on a mullite fibre reinforced metakaolin geopolymer. The fabrication process is shown in Figure 10.8. Mullite fibres were dispersed in water followed by the addition of metakaolin and alkaline solution under vigorous stirring. The slurry was poured into a plastic container which was placed on the shake table for 1 h. The sample was sealed by plastic film and placed in an autoclave at 60°C for 24 h followed by 6 days of curing at ambient conditions.

The filtration test set-up is shown in Figure 10.9. Filtration was conducted using the prepared geopolymer membrane inserted at the bottom of the top flask. A certain volume of tap water was placed in the top flask. The collection flask at the bottom was under vacuum.

The water flux through the membrane is shown in Figure 10.10. It occurs that 300 ml of water passed through the membrane of 36 mm in diameter within 60 min. The flow rate is equivalent to 0.3 m^3/m^2h. However, it was noted that the filtration time increased gradually due to an observed layer of impurities forming on the membrane surface from the tap water used in the experiment. Despite the impurity build-up, the membrane demonstrated effective performance throughout the filtration process. Further research on how to eliminate the impurity build-up through surface treatment is needed.

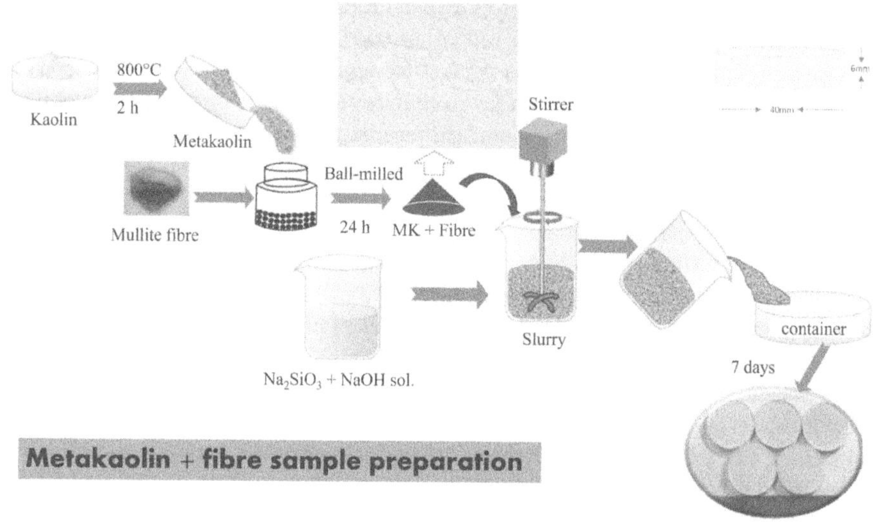

FIGURE 10.8 Preparation of fibre reinforced metakaolin geopolymer filter membrane.

FIGURE 10.9 Test design for geopolymer membrane.

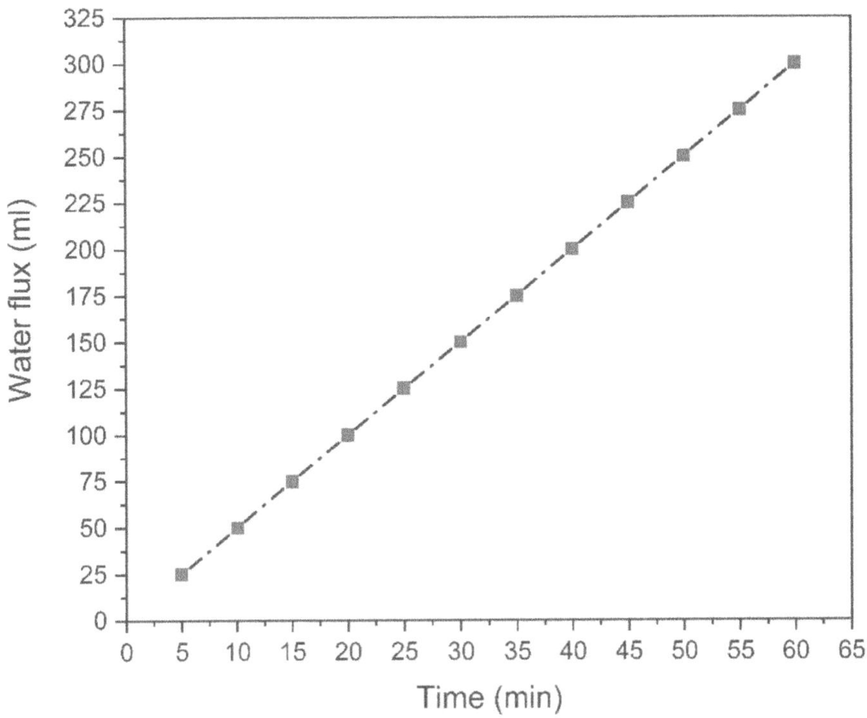

FIGURE 10.10 Water flux through the geopolymer membrane with time.

10.3 THE COMPARISON BETWEEN GEOPOLYMER MEMBRANE AND TRADITIONAL SINTERED CERAMICS

10.3.1 Advantages of Geopolymers Compared to Traditional Sintered Ceramics

The fact that geopolymers require less energy to produce than sintered ceramics is one of their most obvious benefits. In contrast to ceramics, which require a high-temperature sintering process (up to 1400°C), geopolymers are usually synthesised at ambient temperature or at somewhat elevated temperatures (60–90°C). Because of this substantial energy savings, geopolymers are a more environmentally friendly option. For example, studies have demonstrated that the production of geopolymers requires 60% to 80% less energy compared to traditional ceramics, making them more cost-effective, environmentally friendly, and suitable for industrial applications [16]. In addition to lessening the environmental impact, the synthesis of geopolymers from a variety of waste materials, including fly ash, metakaolin, and slag, repurposes industrial by-products that would otherwise contribute to pollution. Because of this, geopolymers are a desirable material for businesses looking to lessen their carbon footprint [16]. On the other hand, traditional ceramics usually use naturally occurring clays that need to be extracted and processed, which has a greater environmental cost. Numerous studies have highlighted the positive effects of geopolymers on the environment, especially how they help advance the ideas of the circular economy by converting trash into useful materials [11].

Compared to conventional ceramics, geopolymers offer more control over microstructural characteristics such as porosity, pore size distribution, and mechanical strength. Geopolymers can be customised for certain filtration applications by varying the proportions of basic materials, activators, and additives [9]. For example, porosity can be controlled, and filtration efficiency can be increased by adding stabilisers like SDS or foaming agents like hydrogen peroxide. However, because of the strict control needed during the high-temperature sintering process, traditional sintered ceramics have less tunability [9]. Geopolymers exhibit excellent chemical resistance, especially in extremely alkaline conditions. Geopolymer materials' aluminosilicate network, which offers resilience against chemical attacks, is the cause of this resistance. According to research, geopolymer membranes are perfect for use in wastewater treatment since they can withstand severe industrial conditions without degrading. In contrast, geopolymers can equal or surpass the high chemical resistance of conventional ceramics in certain chemical conditions [11].

Due to their lower energy consumption and ability to be made from inexpensive and plentiful industrial by-products, geopolymers are typically more affordable than traditional ceramics [16]. For instance, fly ash, a by-product of burning coal, is frequently utilised as a raw material for geopolymers, which significantly lowers the cost of raw materials. However, traditional ceramics have greater production costs since they use more expensive ingredients and energy-intensive procedures [11]. When compared to conventional ceramics, the manufacture of geopolymers results in substantially fewer carbon emissions since they don't require high-temperature processing or the use of carbon-based fuels. According to studies, compared to traditional ceramic production methods, the manufacture of geopolymer materials can reduce CO_2 emissions by as much as 80%. This environmental advantage supports global sustainability objectives, especially for sectors looking to lower their carbon impact [9].

10.3.2 Weaknesses of Geopolymers Compared to Traditional Sintered Ceramics

When compared to conventional sintered ceramics, lower mechanical strength of geopolymers, especially in high-load or high-pressure applications, is one of their main drawbacks. Traditional sintered ceramics can attain much higher values, frequently exceeding 100 MPa, while geopolymers can only reach adequate compressive strength (usually between 10 and 20 MPa). The use of geopolymers in applications where great durability and resistance to mechanical stress are essential, like in structural components subjected to heavy loads or considerable wear and tear, is limited by these differences in mechanical characteristics [9, 11].

Geopolymers are more susceptible to deterioration in acidic circumstances than conventional ceramics despite their remarkable performance in alkaline environments. For example, when exposed to very acidic solutions, fly ash-based geopolymer membranes have demonstrated a propensity to leach heavy metals, which restricts their application in specific filtration procedures involving acidic industrial effluents.

Because of their superior resilience to acidic conditions, traditional sintered ceramics like those made of alumina are better suited for a wider variety of industrial filtration applications [9, 11]. The thermal stability of geopolymers is good, especially when compared to organic membranes, although it is not as good as that of typical ceramics at high temperatures. While sintered ceramics can tolerate temperatures above 1400°C, geopolymer materials can often resist temperatures as high as 800°C. The use of geopolymers in high-temperature industrial applications, including gas filtration in furnaces or kilns, where ceramics perform better, is limited by this disparity in thermal performance [9, 11].

Particularly in water filtration applications, geopolymers have demonstrated problems with long-term durability under continuous operating circumstances. For instance, over time, especially in extremely abrasive settings, certain geopolymer membranes show signs of structural breakdown or pore clogging. Because of their greater mechanical stability and resistance to wear, classic sintered ceramics exhibit less of this constraint and retain their integrity over extended years of use [9, 18]. Variability in material characteristics based on the synthesis process and raw material source is another drawback of geopolymers. Fly ash and other industrial by-products can differ greatly in composition, which can impact the consistency and functionality of geopolymer products. Conventional ceramics tend to have more consistent properties since they are usually made using more standardised raw ingredients and procedures. This is important for applications that need exact performance [11, 12]. The scalability of geopolymer production for large-scale industrial applications is still difficult despite its many benefits. Achieving consistent product quality on an industrial scale is challenging due to the unpredictability of raw materials and the requirement for exact control over the synthesis process. Conventional ceramics are better suited for large-scale production since they have established infrastructure and production methods [11, 18].

10.4 CONCLUSIONS

In summary, geopolymers show great promise as a substitute for conventional sintered ceramics in filtration applications, especially because of their affordable price, environmentally friendly nature, and versatile microstructure. The polymerisation process makes utilising industrial by-products like fly ash and slag possible, reducing production costs and promoting environmental sustainability and waste reduction. Because of their environmentally benign nature, geopolymers are very beneficial in a world where sustainable solutions are becoming increasingly important in industrial processes. Geopolymers provide adaptability in both water and air filtration systems because of their exceptional thermal stability, resistance to a wide range of chemicals, and intrinsic porosity, which can be precisely adjusted to meet specific filtration requirements. The controlled use of activators and additives gives them microstructural flexibility, enabling the construction of membranes with pore diameters appropriate for various filtering applications, from microfiltration to ultrafiltration. In industrial filtering operations, where pollutant removal efficiency and specificity are critical, the capacity to regulate porosity at such an acceptable level is especially advantageous. However, despite their many benefits, a few issues must be resolved to fully utilise geopolymers' potential infiltration. While geopolymers are sufficient for a wide range of applications, their mechanical strength is somehow inferior to sintered ceramics, especially in high-stress or impact situations. Furthermore, geopolymers have difficulties in acidic settings because alkali element leaching can weaken their structural stability. These drawbacks emphasise the necessity of more study to improve their mechanical characteristics and create plans for increased resilience in challenging chemical environments.

Future research should concentrate on reinforcing geopolymer structures to improve their mechanical robustness, maybe using hybrid techniques that include fibre or nanoparticle reinforcements. Expanding the use of geopolymers in industries with demanding chemical requirements may also be possible by looking at alternate curing techniques or changing the chemical makeup of the geopolymers to increase their resistance to acid. Furthermore, formulations that optimise porosity and structural integrity may be developed by comprehending how pore-forming agents and geopolymer matrices interact. Geopolymers have great potential as an effective, flexible, and ecological substitute for traditional ceramics in filtration applications. As material composition and processing techniques continue to advance, geopolymers could play an increasingly vital role in industrial filtration, fulfilling the performance requirements of various filtration systems while also supporting environmental sustainability goals.

REFERENCES

1. Davidovits, J., *Geopolymer chemistry and applications.* 2008: Geopolymer Institute.
2. Davidovits, J., *Geopolymers and geopolymeric materials.* Journal of Thermal Analysis, 1989. **35**: p. 429–441.
3. Lecomte, I., et al., *Synthesis and characterisation of new inorganic polymeric composites based on kaolin or white clay and on ground-granulated blast furnace slag.* Journal of Materials Research, 2003. **18**(11): p. 2571–2579.
4. Palmero, P., et al., *Geopolymer technology for application-oriented dense and lightened materials. Elaboration and characterisation.* Ceramics International, 2015. **41**(10, Part A): p. 12967–12979.
5. Hemra, K. and P. Aungkavattana, *Effect of cordierite addition on compressive strength and thermal stability of metakaolin based geopolymer.* Advanced Powder Technology, 2016. **27**(3): p. 1021–1026.
6. da Silva Biron, D., V. Dos Santos, and M. Zeni, *Ceramic membranes applied in separation processes.* 2017: Springer.
7. Jämstorp, E., et al., *Mechanically strong geopolymers offer new possibilities in treatment of chronic pain.* Journal of Controlled Release, 2010. **146**(3): p. 370–377.
8. Ma, Y., J. Hu, and G. Ye, *The pore structure and permeability of alkali activated fly ash.* Fuel, 2013. **104**: p. 771–780.
9. Bai, C. and P. Colombo, *Processing, properties and applications of highly porous geopolymers: A review.* Ceramics International, 2018. **44**(14): p. 16103–16118.
10. Abdullayev, A., et al., *Materials and applications for low-cost ceramic membranes.* Membranes, 2019. **9**(9): p. 105.
11. Mohd Mortar, N.A., et al., *Geopolymer ceramic application: A review on mix design, properties and reinforcement enhancement.* Materials, 2022. **15**(21): p. 7567.
12. Goswami, K.P. and G. Pugazhenthi, *Credibility of polymeric and ceramic membrane filtration in the removal of bacteria and virus from water: A review.* Journal of Environmental Management, 2020. **268**: p. 110583.

13. Ahmad, R., et al., *Processing and properties of polymer filled geopolymer ceramics fabricated via powder metallurgy method: A review.* Reviews on Advanced Materials Science, 2016. **44**(1): p. 26–32.
14. Bolong, N., I. Saad, and S. Arshad, *Effect of sodium hydroxide and mould shape in geopolymer fabrication made with synthesised local clay for water filtration.* Journal of Applied Membrane Science & Technology, 2016. **18**(1).
15. Goswami, K.P., K. Pakshirajan, and G. Pugazhenthi, *Process intensification through waste fly ash conversion and application as ceramic membranes: A review.* Science of The Total Environment, 2022. **808**: p. 151968.
16. Sadiq, M., et al., *Geopolymerization: A promising technique for membrane synthesis.* Materials Research Express, 2021. **8**(11): p. 112002.
17. Cheng, Y., et al., *A review on optimistic development of polymeric nanocomposite membrane on environmental remediation.* Chemosphere, 2023. **315**: p. 137706.
18. Nurlina, N., et al., *A review of geopolymer membrane for water treatment.* Applied Clay Science, 2024. **251**: p. 107301.
19. Rani, S.L.S. and R.V. Kumar, *Insights on applications of low-cost ceramic membranes in wastewater treatment: A mini-review.* Case Studies in Chemical and Environmental Engineering, 2021. **4**: p. 100149.
20. Arumugham, T., et al., *Recent developments in porous ceramic membranes for wastewater treatment and desalination: A review.* Journal of Environmental Management, 2021. **293**: p. 112925.
21. Valappil, R.S.K., N. Ghasem, and M. Al-Marzouqi, *Current and future trends in polymer membrane-based gas separation technology: A comprehensive review.* Journal of Industrial and Engineering Chemistry, 2021. **98**: p. 103–129.
22. Lorente-Ayza, M.-M., et al., *Comparison of porosity assessment techniques for low-cost ceramic membranes.* boletín de la sociedad española de cerámica y vidrio, 2017. **56**(1): p. 29–38.
23. Xu, M.-X., et al., *Preparation and characterisation of a self-supporting inorganic membrane based on metakaolin-based geopolymers.* Applied Clay Science, 2015. **115**: p. 254–259.
24. Xu, M.-X., et al., *Preparation of geopolymer inorganic membrane and purification of pulp-papermaking green liquor.* Applied Clay Science, 2019. **168**: p. 269–275.
25. Naveed, A., et al., *Synthesis and characterisation of fly ash based geopolymeric membrane for produced water treatment.* Desalination and Water Treatment, 2019. **161**: p. 126–131.
26. Naveed, A., et al., *Porosity control of self-supported geopolymeric membrane through hydrogen peroxide and starch additives.* Desalination and Water Treatment, 2019. **152**: p. 11–15.
27. Korat, L. and V. Ducman, *The influence of the stabilising agent SDS on porosity development in alkali-activated fly-ash based foams.* Cement and Concrete Composites, 2017. **80**: p. 168–174.
28. Song, Y., et al., *Chitosan-modified geopolymer sub-microparticles reinforced multifunctional membrane for enhanced removal of multiple contaminants in water.* Journal of Membrane Science, 2022. **658**: p. 120704.
29. Zhang, J., et al., *Facile fabrication of a low-cost and environmentally friendly inorganic-organic composite membrane for aquatic dye removal.* Journal of Environmental Management, 2020. **256**: p. 109969.

Index

I

L

M

N

O

P

R

S

T

W

X

For Product Safety Concerns and Information please contact our EU representative GPSR@taylorandfrancis.com
Taylor & Francis Verlag GmbH, Kaufingerstraße 24, 80331 München, Germany

www.ingramcontent.com/pod-product-compliance
Ingram Content Group UK Ltd.
Pitfield, Milton Keynes, MK11 3LW, UK
UKHW020203200926
475702UK00004B/595
* 9 7 8 1 0 3 2 8 7 9 7 3 4 *